MAB Technical Notes 7

Titles in this series:

1. *The Sahel: ecological approaches to land use*
2. *Mediterranean forests and maquis: ecology, conservation and management*
3. *Human population problems in the biosphere: some research strategies and designs*
4. *Dynamic changes in terrestrial ecosystems: patterns of change, techniques for study and applications to management*
5. *Guidelines for field studies in environmental perception*
6. *Development of arid and semi-arid lands: obstacles and prospects*
7. *Map of the world distribution of arid regions*

Map of the world distribution of arid regions

Explanatory note

Geographisches Institut
der Universität Kiel
Neue Universität

unesco

Launched by Unesco in 1970, the intergovernmental Programme on Man and the Biosphere (MAB) aims to develop within the natural and social sciences a basis for the rational use and conservation of the resources of the biosphere and for the improvement of the relationship between man and the environment. To achieve these objectives, the MAB Programme has adopted an integrated ecological approach for its research and training activities, centred around fourteen major international themes and designed for the solution of concrete management problems in the different types of ecosystems.

Published in 1979 by the United Nations
Educational, Scientific and Cultural Organization
7 Place de Fontenoy, 75700 Paris
Printed by Union Typographique,
Villeneuve-Saint-Georges

ISBN 92-3-101484-6

*Carte de la répartition mondiale
des régions arides :* 92-3-201484-X

© Unesco 1977
Printed in France

Preface

From 1951 to 1964 Unesco undertook a world-wide research programme on arid zones, in order to encourage and advance the study of the problems of these regions. This programme included publication of thirty volumes in the 'Arid Zone Research' series, and the development of important research and training institutions.

Unesco's subsequent research and study activities on arid and semi-arid zones have taken place within the framework of the Programme on Man and the Biosphere (MAB) and of the International Hydrological Programme (IHP). Two of the fourteen main themes of the MAB Programme are directly related to arid land problems: MAB Project 3 on the impact of human activities and land use practices on grazing lands, including those in arid and semi-arid zones, and Project 4 on the impact of human activities, especially irrigation, on these ecosystems. An integrated research project on arid land management (IPAL) is now in its operational phase in several countries and the expected results will be important for rational development.

This long tradition naturally led Unesco to associate itself closely with the preparation of the United Nations Conference on Desertification (Nairobi, 29 August to 9 September 1977) and with the preparation of the necessary working documents. In addition to several MAB Technical Notes, on the Sahel (MAB Technical Notes 1), Mediterranean forests and maquis (MAB Technical Notes 2), the obstacles and prospects for development of arid lands (MAB Technical Notes 6) and irrigation in developing countries (MAB Technical Notes 8), and six case studies on desertification throughout the world, Unesco has prepared a new *Map of the World Distribution of Arid Regions*. The last map on this subject published by Unesco was prepared by Meigs in 1952. The knowledge gained during the last two decades about climates, soils and vegetation in these regions, as well as methods for classifying aridity, was used to better map the arid and semi-arid lands and to provide the basis for useful comparisons between different parts of the world. The map is accompanied by a summary of basic information on climate, vegetation and land-use patterns of the major arid regions.

This map and the accompanying document were prepared in close collaboration with specialists from FAO and WMO, as well as from UNEP. Unesco wishes to express its profound gratitude to all those who participated in this endeavour. The compilation of the map was entrusted to the Laboratoire de Cartographie Thématique of the Centre National de la Recherche Scientifique in Paris, J. Mallet and R. Ghirardi being responsible for this task; the regional summaries were prepared by F. Blasco and P. Legris (of the Institut de la Carte du Tapis Végétal, Université Paul Sabatier, Toulouse) and by E. A. Fitzpatrick (Geography Department, University of New South Wales, Australia). Unesco wishes to thank these authors, as well as researchers in many countries, whose comments were invaluable in finalizing this publication.

The designations employed and the delimitations of frontiers on this map and in the accompanying text do not imply the expression of any opinion whatsoever on the part of Unesco concerning the legal or constitutional status of any country.

Contents

Introduction 9
   Preparing the map 10
   Descriptive text 12

Regional presentation 15
   Countries of Mediterranean Europe 15
   Maghrib countries and Libyan Arab Jamahiriya 19
   Countries of the Near and Middle East 21
   Central Asia 26
   Indian sub-continent 29
   Australia 32
   Sahara, Sahelian and Sudanese zones 38
   Southern, south-western Africa and Madagascar 41
   North America 41
   South America 45

Bibliography 49

Introduction

Nearly half the countries of the world face problems of aridity, and for more than two decades the international community has shown a keen interest in the problems of arid and semi-arid zones. From 1951 to 1964, Unesco conducted a world programme of research to encourage and advance study of the problems of these regions. In the last decade, Unesco has also participated in the preparation of several small-scale thematic maps which synthesize current knowledge of the natural resources of the arid and semi-arid zones. These include: the world soil map, the bioclimatic map of the Mediterranean zone, vegetation maps of the Mediterranean region, of Africa and of South America, and continental geological maps. The recent droughts which struck huge areas of Africa and Asia dramatically highlighted the need for urgent solutions in these zones and the complexity of the obstacles to their rational development.

Unesco has naturally been associated with this world-wide effort, in particular through the preparation of several MAB Technical Notes on arid zone problems. MAB Technical Notes 6, *Development of Arid and Semi-arid Lands: Obstacles and Prospects*, was prepared in this spirit at the request of the Committee on Science and Technology for Development. Unesco also contributed actively to the preparation of the United Nations Conference on Desertification (Nairobi, 29 August to 9 September 1977). This conference took stock of a long list of problems concerning both the extension of desert conditions beyond the present boundaries of deserts, and increasing aridity within arid and semi-arid regions, which is leading to a collapse in productivity and living standards. Such an analysis needed to call upon objective knowledge of the causes and manifestations of aridity and to be based on as accurate as possible a delimitation of the distribution of the world's arid and semi-arid zones.

The geographical distribution of arid zones obviously depends on how they are defined and should be shown cartographically. A world map based on Thornthwaite's index was prepared at Unesco's request by P. Meigs in 1952. This map showed the distribution of arid climatic zones on two sheets, at a scale of 1 : 25,000,000. The only one of its kind, and relatively detailed despite its small scale, this map has hitherto served as a reference point for all those interested in the study of the arid and semi-arid zones. However, in the last two decades, important new data have been gathered, not only on climate but also on biology. A denser meteorological network, progress in detailed climatology and numerous field studies of soils and vegetation, have contributed to a better knowledge of the arid regions. At the same time, a better understanding of the complex relationships between climate, soils and types of plant cover was achieved. A new map thus became necessary, which would not only more accurately show the limits of the major climatic zones but also, by including as much biological information as possible, have a strong bioclimatic emphasis. This task had to be undertaken rapidly, in view of the preparation for the United Nations Conference on Desertification of a world map of desertification. This map, established by FAO, Unesco and WMO, and published by UNEP for the Conference (UNEP-FAO-Unesco-WMO, 1977) needed a delimitation of arid, semi-arid and sub-humid regions.

Starting from this regional delimitation, it was possible to prepare the new Unesco map of the distribution of arid regions which is presented here. This map was drawn by J. Mallet and R. Ghirardi of the Laboratoire de Cartographie Thématique of the Centre National de la Recherche Scientifique in Paris. At different stages in preparing the map, the advice of numerous specialists was sought, in order to include as much biological and bioclimatic information as possible.

Preparing the map

The scale is the same as that of the Meigs map, since it allows the world to be shown in a convenient format on a single sheet. The projection of the present map is, on the other hand, very different. It is based on that of the world map in sixteen sheets at a scale of 1 : 5,000,000 published by the American Geographical Society of New York. The outline map at 1 : 25,000,000 was obtained by photographic reduction of the 1 : 5,000,000 map, and by contracting the oceans. The projection is unique for the Americas (bipolar oblique conformal projection); elsewhere, Miller's stereographic system, flattened in three conformal zones, has been used. This system deforms the various regions very little and represents surface areas adequately. It should be noted that this projection was adopted for all the 1 : 25,000,000 maps prepared for the United Nations Conference on Desertification.

The delimitation of arid and semi-arid regions is based partly on aridity indices, and partly on consideration of all available data on soil, relief and vegetation.

The degree of bioclimatic aridity depends on the relative amounts of water gained from rainfall and lost by evaporation and transpiration: aridity rises as precipitation decreases and as evaporation increases. Thus the values of the ratio P/ETP (in which P is the mean value of annual precipitation, and ETP is the mean annual potential evapotranspiration) have been used here to delimit arid and semi-arid regions. The ratio P/ETP was used in preference to the difference $P - ETP$, which refers rather to the amount of water available and which can be the same for many different climates (for example, $P - ETP = 400$ can result from 1,000 - 600 or 800 - 400, or 600 - 200, etc.). On the other hand, in arid and semi-arid areas, the ratio P/ETP expresses the degree of aridity better, because it gives the same value for all climates in which the potential water loss is proportionally the same in relation to rainfall. Aridity increases as values of this ratio decline. In addition, this ratio is biologically accurate in climates with highly contrasted seasons, since it represents well the ratio ETR/ETM (ETR = real evapotranspiration of a soil-plant system; ETM = maximum evapotranspiration in the absence of a water constraint), which largely determines vegetative dry-matter production. In these areas, annual precipitation (P) is a satisfactory proxy for ETR, and potential evapotranspiration (ETP) is very close to ETM, being its upper limit.

The ratio P/ETP was also used by FAO in its study of desertification risk (Riquier and Rossetti 1976), and it was desirable to use the same index to delimit bioclimates in order to allow simple comparisons between the different maps.

For the purpose of the map, the ratio P/ETP was calculated by D. Henning of the Meteorologisches Institut of the University of Bonn, using data for 1,600 stations provided by WMO. Calculating P did not pose any particular problems, whereas there are several ways of calculating ETP, each with advantages and disadvantages depending on the climate concerned. Penman's formula was used here; it includes solar radiation, atmospheric humidity and wind (very important in arid and semi-arid zones, because of its drying power on the air). It is true that the delimitations made using this formula are to a certain degree arbitrary, as are those given by the various other climatic indices that were proposed. Aridity indices such as those proposed by Penman, Budyko and others are also mathematically related to one another and are thus to a certain degree interchangeable (Hare, 1977). The great advantage of Penman's formula is that it has been used in numerous biological and physical studies of climate, the results of which have been widely diffused. In addition, it is today considered more satisfactory than the formula used in the Meigs map.

Mean annual values of the ratio P/ETP were put on to the chosen outline map before its reduction to the final scale, that is to say at its initial scale of 1 : 5,000,000. The limits of bioclimatic regions were then drawn by interpolation, taking into account, for the regions where there were few climatic data, the information given on the FAO/Unesco *Soil Map of the World* and maps of vegetation, all at the same scale. In addition, where mappable information was not conclusive, great attention was given to field observations, using the expertise of specialists on different parts of the world. It is clear that the limits thus drawn, however carefully, are somewhat arbitrary, in particular in areas where information is scarce. Since with few exceptions, such as along a sharp change in altitude or a coastline, phenomena on the ground vary continuously and progressively, it is difficult to put a clear limit to the features shown. This is an extremely important obstacle to any bioclimatic mapping. To this problem is added a double spatial and temporal variability in the values of the main climatic components used here. The map was established on the basis of the most widely available data, which are means calculated over a very great number of years. Use of these means conceals real variations, the consequences of which may be important for the natural vegetation, agriculture or pastoralism, all the more since year to year variation is especially great in arid and semi-arid regions.

Four main classes or degrees of aridity have been delimited, corresponding to the major geographic categories generally used by climatologists and biologists.

The *hyper-arid zone* ($P/ETP < 0.03$) is shown on the map by single colours bordered by a continuous flagged black line. It corresponds to real desert climates, with very low and irregular rain which may fall in any season. These regions have almost no perennial vegetation, except some bushes in river beds; annual plants can grow in good years. Agriculture and grazing are generally impossible, except in cases. Interannual variability of rainfall can reach 100 per cent.

The *arid zone* ($0.03 < P/ETP < 0.20$) is shown on the map by single colours bordered by a continuous grey line. The vegetation of this zone is scattered, and includes, according to the region, bushes and small woody, succulent, thorny or leafless shrubs. Very light pastoral use is possible, but no rainfed agriculture. These regions are characterized by annual rainfall of 80-150 mm and 200-350 mm; interannual rainfall variability is 50 to 100 per cent.

The *semi-arid zone* ($0.20 < P/ETP < 0.50$) is shown on the map by colours streaked with white and bordered by a dashed grey line. This is a steppe zone, with some savannahs and tropical scrub. These are sometimes good grazing areas and rainfed agriculture is possible, although the harvest is often irregular due to great rainfall variability. Mean annual rainfall in this zone varies between 300-400 mm and 700 or even 800 mm in summer rainfall regimes, and between 200-250 and 450-500 mm in winter rainfall regimes, at Mediterranean and tropical latitudes. Interannual rainfall variability is between 25 and 50 per cent.

The *sub-humid zone* ($0.50 < P/ETP < 0.75$) is shown on the map by colours overlaid with white diamonds, not bordered towards wetter zones because the transitions are extremely variable. This zone includes mainly certain types of tropical savannah, maquis and chaparral in Mediterranean climates, steppes on chernozem soils, etc. Agriculture is the normal use. Interannual rainfall variability is less than 25 per cent. The sub-humid zone was not shown on the Meigs map, but it seemed necessary to introduce it because desertification as a result of soil and vegetation degradation also occurs in this zone. Only quite large areas have been included, and more localized areas of desertification, for example in Yugoslavia or in New Caledonia, have been omitted.

Finally, it should be noted that drought may also occur in parts of the world here considered to be humid; they are not taken into account in this map in order not to extend unreasonably the area where aridity is the main constraint.

In addition to the four aridity classes defined above, it was necessary to take account of temperature criteria to introduce new subdivisions. Temperature and its annual variations are, with precipitation, an important influence on plant production.

The use of temperature criteria is reflected in subdivisions based in the first place on the mean temperature, in °C, of the coldest month of the year. The four classes defined are: (a) warm winter, mean temperature of the coldest month 20 to 30° C; (b) mild winter, mean temperature of the coldest month 10 to 20° C; (c) cool winter, mean temperature of the coldest month 0 to 10° C, and (d) cold winter, mean temperature of the coldest month $< 0°$ C.

These four classes are shown by four colours of decreasing intensity: red-ochre, orange, yellow and green. These temperature classes are in turn subdivided according to the mean temperature of the hottest month of the year, the limiting values being 10°, 20° and 30° C. These subdivisions are shown by three tones (dark, medium and light) of each of the four basic winter colours. The use of these two series of temperature criteria enables the mean annual temperature range, which varies according to continentality and latitude, to be shown. Thus the map colours correspond to temperature variations, and not to the major aridity classes defined above.

These temperature data were already taken into account in the Meigs map, but consideration has also been given here to the position of the rainy period in relation to seasonal temperatures, since this has a unique biological importance. One of the innovations of this map as compared to that of Meigs is thus to show the length of dry periods and the rainfall pattern. The first is shown by the size of a small circle at the site of a certain number of climatological stations, the second by the colour of this circle.

On the map there are circles of six sizes, representing six different lengths of drought period, determined by the number of dry months. For the purpose of the map, any month with less than 30 mm has been arbitrarily considered dry. This simple definition, used by Aubréville (1949) among others, gives, in these dry climates, results which differ little from those based on the formula $P < 2\,t$ of Bagnouls and Gaussen (1957), used by Walter and Lieth (1960). Data from 4,000 stations gathered by Walter and Lieth in their *Climadiagram Atlas* have been analysed, the dry periods calculated according to this method, and information on about 1,000 stations has been included on the map.

Six rainfall regimes are shown and represented by six circle colours. They correspond to two regimes with predominantly dry summers, two regimes with predominantly dry winters, and two transition regimes. The position of the dry season is important for growth and productivity of the vegetation. The small scale of the planisphere and the nature of the map have led to some simplification; in particular, only dominant seasonal patterns are taken into account, and these do not always correspond to the real complexity of combinations of temperature and rainfall patterns.

The information on this map has been considerably increased and brought up to date in relation to that of Meigs. Forty-four cartographic categories are distinguished according to climatic conditions, and in addition they carry information on drought for about 1,000 stations. The map can be read more easily than that of Meigs, since all the information chosen is shown graphically and by colours, and not by symbols. The shades representing the different cartographic categories were chosen in order to give a gradient of colours and shades, thus emphasizing the absence of clear dividing lines on the ground, as a result of the slow transformation of biological situations and the great variability of climatic phenomena in these areas.

On the other hand, the shortcomings of the map are clear. The scale chosen makes inevitable a high degree of generalization, and the limits of zones are only indicative of complex and changing field situations. Attention has already been drawn to the fact that all climatic aridity indices are more or less arbitrary and portray biological conditions only approximately, especially when they are used at a planetary level. In addition, the indices only use mean values; interannual variability of climatic phenomena is not taken into account, although, as already pointed out, it is of fundamental importance in arid and semi-arid zones. Thus, an area which regularly receives 200 mm of winter rainfall could be considered suitable for rainfed agriculture; but if variability is greater than 40 per cent, with the same rainfall, no rainfed cultivation is possible. Even if mean rainfall is the same in both cases, from an agronomic point of view, aridity is considerably more marked in the area of greater variability.

In order to overcome the shortcomings resulting from sole reliance on climatic indices, whatever their merits, account was taken in preparing the map of all available information, from topography, soil and vegetation maps, and from direct observation. The map thus seeks a synthesis and an integrated approach to aridity, in order to give as objective as possible an overview of the phenomenon. It shows clearly the scale and diversity of these conditions and makes possible some comparisons between different parts of the world. The map will thus be of use to planners, decision-makers, geographers, and more generally to all those concerned with potential land use. It provides agronomists and livestock specialists with a basic tool for comparisons and experiments between different regions. Finally, it should be a valuable aid for teaching and research.

Descriptive text

It was considered essential to accompany the map with a document giving a general description of the regions shown. This document was prepared by P. Legris and F. Blasco, of the Institut de la Carte du Tapis Végétal at the Université Paul Sabatier at Toulouse, in consultation with specialists in several countries. The Australia section was written by E. A. Fitzpatrick, Geography Department, University of New South Wales.

This document could have been prepared in two different ways. It could have followed the map closely, and listed the bioclimatic features of the four main aridity zones: hyper-arid, arid, semi-arid and sub-humid. But a presentation by subdivisions of the aridity zones would have inevitably led to repetition. On the other hand, it was possible to follow a more regional and more geographical approach, by describing separately the different arid regions of the various continents. In the end, this second approach was adopted, because it facilitated the task of those (including non-specialists) with a particular interest in a specific region.

The categories used are thus essentially geographical regions. However, for convenience, some groupings are defined by the names of the main countries concerned. It is clear that the ecological limits of these groupings do not necessarily coincide with the political boundaries of the States concerned.

In each region, climatic characteristics and dominant vegetation patterns are described, mentioning, where possible, the most characteristic or most useful plants. Hydrology, main soil types and main land uses complete the information given.

To illustrate bioclimates, ombrothermic diagrams, drawn according to the methods of Bagnouls and Gaussen (1957) and Walter and Lieth (1960), are used. These diagrams follow a convention; a scale convention (P mm $= 2t$ in °C) illustrates graphically the biologically dry period of the year. The graphs are shaded to read more easily, but this does not of course show intensity of aridity. Based on monthly means, these diagrams cannot be considered the rain and temperature patterns of a particular year; the points where the curves cross, showing the limits of wet and dry seasons, are only indicative. However, this type of diagram is widely used in small-scale work by ecologists and biogeographers, and it may thus be used for general comparisons.

A great many publications are now available on arid zones. Reference has only been made to atlases and maps at various scales, some of which were used for this document. Among the general works, those with a very large list of references have been chosen, as have bibliographies. In each regional study a few basic works are cited.

This descriptive text is not exhaustive; it could not be, given its length and the scale of the map. The advice of a number of specialists was sought on this text and their comments have been incorporated in it. Additional comments made by other specialists on reading this Technical Note will provide material for future preparation of a more complete version.

Regional presentation

The main types of arid climate and land use are discussed in the following order: Mediterranean Europe; Maghrib and Libyan Arab Jamahiriya; Near and Middle East; Central Asia; Indian Sub-Continent; Australia; Sahara; Sahelian and Sudanese Zones; Africa and Madagascar; Americas.

Countries of Mediterranean Europe

This chapter deals with: the Iberian Peninsula, southern France, Italy and Greece.

Iberian Peninsula

Figure 1 includes a schematic representation of the climate at three stations in the Iberian Peninsula.

Climates

A simple distinction can be made according to the rainfall regime:

Those with winter rain and very pronounced summer drought: Southern Portugal (Faro, Fig. 1) and Andalusia (Alicante, Fig. 1). The rains are sometimes delayed until spring, but there are usually four or five dry months. This is olive and cork oak country and cotton cultivation is increasing.

Those with two rainfall maxima, in spring and autumn. Eastern Spain and Castilla have this double dry period. It also occurs in the Balearic Islands.

The most severe drought occurs on eastern slopes and on the east coast, especially in the Almeria region (Fig. 1) where there are ten or even eleven dry months. In Castilla, the dry season is about six months long (see especially Montero de Burgos and Gonzales Rebollar, 1974). Land use varies considerably according to local irrigation possibilities.

Natural vegetation

Although the vegetation is as a whole extremely degraded, it does, however, illustrate the different degrees of aridity shown on the map.

In Eastern Andalusia, the driest region of Spain, a *Stipa* pseudo-steppe is common. A vegetation type close to the climax is a xerophytic scrub of which *Gymnosporia senegalensis, Periploca laevigata, Tetraclinis articulata* and *Salsola webii* are the most characteristic species. Other plants which tolerate dry conditions are also found, such as *Anabasis articulata, Haloxylon articulatum, Launaea arborescens, Lycium intricatum, Withania frutescens* and *Ziziphus lotus*. The work of Freitag (1971) concerns particularly these regions. The date palm (*Phoenix dactylifera*) and various tropical crops (sugar-cane and avocado for example) are grown successfully as long as there is sufficient soil moisture.

Aridity declines gradually towards the north; Aleppo pine becomes dominant on the eastern slopes near the coast and in the Balearic Islands. The even less arid regions of the Ebro (Hijar, Almunia de Dona Godina) and of the Duero (Zamora) have a climax vegetation characterized by *Quercus coccifera* and *Rhamnus lycioides*. These areas are still too dry for Holm oak (*Quercus ilex*) to grow well ('infrailicine ground'). In these regions very strong human and biotic influences are shown by the presence of such species as *Rosmarinus officinalis, Thymus vulgaris, Erica multiflora, Helianthemum racemosum,* and *Cistus* sp. (*C. libanotis*), etc.

In the Ebro valley around Zaragoza, there is a phytogeographical enclave which is interesting for its stands of *Juniperus thurifera*, a species fairly characteristic of climates with harsh winters and arid

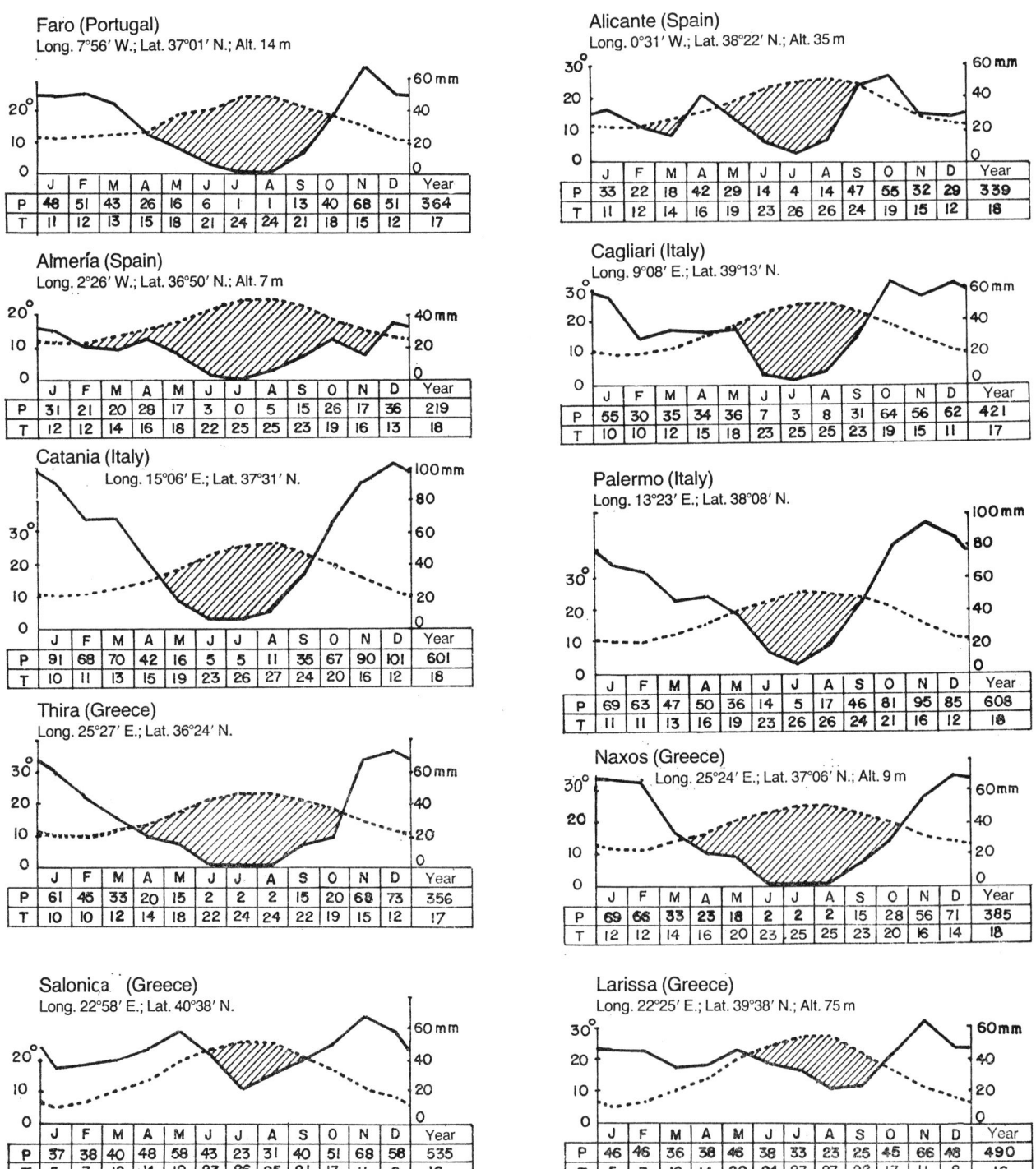

FIG. 1. Schematic representation of the climate at several stations in Greece, Italy, Portugal and Spain. (P = precipitation; T = temperature).

summers. A steppe-like vegetation with *Lygeum spartum* is very characteristic of this area (Durrieu, 1967). A careful distinction should be made between climates in such cases. The distribution of these stands is quite well delimited in the field by juniper, *Rhamnus lycioides, Genista scorpius* and by the species associated with degraded stages, such as artemisia, *Salsola vermiculata,* thyme, *Asphodelus fistulosus,* etc. These stages are the *tomillares.*

Land use

Information on the soils and land use of the dry regions of Spain is in Roquero de Laburu (1964) and Allue Andrade and Navarro Garnica (1970).

In the driest parts of the country (Almeria and Murcia) there are mainly degraded rangelands with rainfed cereal cultivation. Irrigation is only found in the Segura valley and in the lower basins of the Guadalentin and Vinalapo. Large orchards are found west of the coastal lake 'Mar Menor', practically as far as Murcia.

Further north, in Aragon, the reputably dry areas of Hijar, Almunia de Dona Godina and Alfaro near the Ebro contain important vineyards and large areas of cereals. Elsewhere, all the lowlands are developed and irrigated. This is not so further west in the Duero valley (Zamora) where there is little irrigation and most of the land is used for dry cereal farming, with the exception of some areas which by tradition are devoted to vines.

Until now, the species most used for reafforestation of dry habitats is the Aleppo pine, which is a remarkable pioneer species. Other pines have not been used in Spain except for another indigenous species, *Pinus pinea,* of which good stands are found in the Huelva region. *Cupressus* (especially *C. arizonica*), *Eucalyptus camaldulensis* and *E. gomphocephala* have been introduced in Spain, but still on a small scale.

For the last twenty years, *Opuntia inermis* has been cultivated on about 10,000 ha. The purpose was to control erosion and to provide a summer feed supplement for livestock. The results have not been encouraging. This is true also for *Agave gigantea*.

France

In France, the only semi-arid areas are the degraded garigue and *tomillares* of the eastern Corbières and the Clape (400-500 mm mean annual rainfall; a sparse vegetation of *Brachypodium ramosum, Thymus vulgaris, Rosmarinus officinalis, Juniperus oxycedrus,* etc.), and the limestone areas of Provence (the Marseilles region and the Crau). Small regions with sub-humid climates exist in Provence, and on the east coast and extreme south of Corsica (in the Bonifacio area). Using the vegetation map of Corsica (Dupias *et al.,* 1965), it is easy to establish a simple relationship between climatic type and the dominant vegetation of these areas.

Maquis is the characteristic plant formation. It is a dense, low (8-9 m) forest, made up of sclerophyll, evergreen plants. This typical Mediterranean flora includes in particular lentisk (*Pistacia lentiscus*), myrtle (*Myrtus communis*), wild olive (*Olea europaea* var. *oleaster*), *Erica arborea, Cistus monspeliensis, Arbutus unedo, Calycotome villosa, Genista corsica, Phillyrea angustifolia, Daphne gnidium.* In addition, in the driest areas of southern and eastern Corsica, azonal stands of *Juniperus phoenicea* and *Helianthemum halimifolium* are found. Repeated fires have considerably reduced the area of maquis, of which the best examples are now found around the Gulf of Porto Vecchio. Elsewhere, it has been replaced by a lower and less dense garigue which covers large areas of this dry part of Corsica. Important commercial forests of cork oak (*Quercus suber*) are also found in the south of the island. The climate and deep siliceous soils suit it well.

Crops occupy only small areas, essentially in the form of gardens. Vineyards and dry grasslands are notable.

Italy

Climates

Using the aridity criteria of Bagnouls and Gaussen (1957), some authors consider that Italy does not have truly arid regions. De Philippis (1970) thinks that the least rainy parts of the country should be designated as 'semi-arid'. These are the coastal areas of Sardinia (Cagliari, Fig. 1), of Sicily (Catania and Palermo, Fig. 1), and the south of Calabria to Taranto.

The rainfall regime in all these areas is characterized by a single rainfall maximum in winter, and a pronounced summer dry period lasting about five months (May to August-September). Mean annual rainfall is usually between 400 and 600 mm. In the lowlands, the temperatures vary little from one station to another. Winters are cool (December and January average between 10° and 12° C), and summers are hot (25.5° to 28° C in July-August).

The variability of temperatures and rainfall from

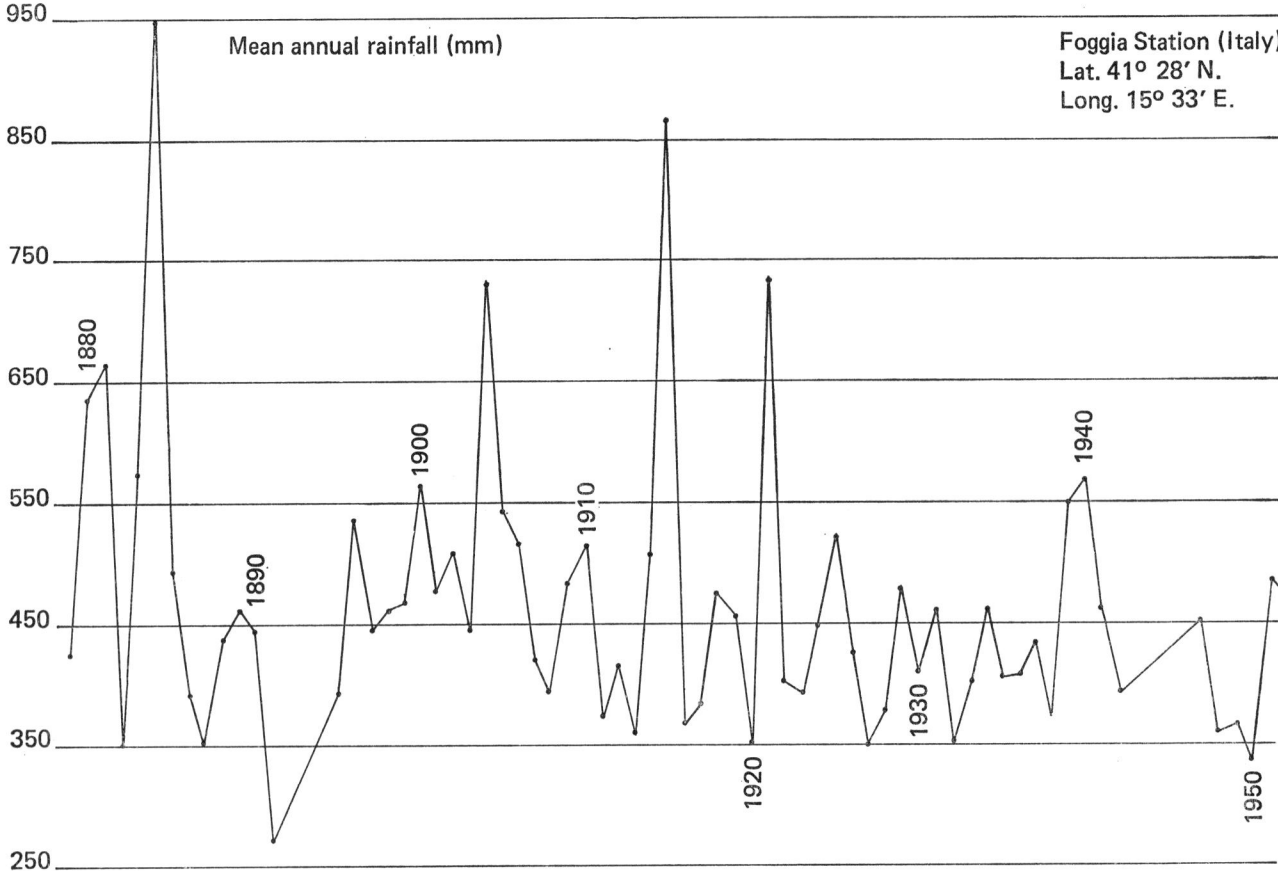

FIG. 2. Interannual rainfall variability at the station of Foggia (Italy). (de Philippis, 1970).

year to year at the same station is not yet adequately known; this variability is shown in Figure 2 for the station of Foggia, where the mean annual rainfall is about 470 mm (de Philippis, 1970). Exceptionally low rainfall, of about 350 mm or even less, is possible whilst very high rainfall over 850 mm has also been registered. This is not surprising in the Mediterranean. It is still difficult to assess accurately the effects of the large and unpredictable variations.

Natural vegetation and land use

Soil types are very varied, as a result of different types of parent rock (Archean rock dominant in Sardinia, calcareous rock on the south slopes of the Sicilian hills, volcanic rocks around Etna). Elsewhere, erosion has considerably influenced soil types.

The natural vegetation which is the most thermophilous and drought tolerant includes several remarkable species such as carob (*Ceratonia siliqua*), wild olive (*Olea europaea* var. *oleaster*), lentisk (*Pistacia lentiscus*), dwarf fan palm (*Chamaerops humilis*) and several Cistus. On the whole, the commonest vegetation type of the Italian arid or semi-arid climates is a fairly open garigue whose floristic composition reflects quite well the soil type: (a) on soils formed over calcareous rock, the characteristic plants include *Rosmarinus officinalis, Thymus capitatus, Poterium spinosum,* etc.; (b) on siliceous soils, Cistus, *Helychrysum* sp., *Lavandula stoechas* are abundant. The commonest natural tree species, especially on limestone, is *Pinus halepensis*. This pine, which tolerates drought, has the additional advantage of regenerating itself abundantly and naturally after a fire.

In Sardinia, cork oaks cover large areas of acid soils. The production varies considerably from one climatic region to another (200 to 500 kg/ha/year of fresh cork). On average, the bark is stripped every nine years.

Holm oak is found almost everywhere in Italy, including the driest regions. These are not trees but bushes resulting from vegetative regeneration. Clearcutting takes place on average every twenty years. Advantage is taken of the cutting to extract the roots of *Erica arborea* used for making pipes. See also Tomaselli (1973) and Tomaselli *et al.* (1973).

Agriculture has a very modest role in these poor regions. There are some irrigated citrus groves. The unirrigated land is planted with olives, vines and cereal crops. The least favoured regions, which are not suitable for cultivation, are given over to uncontrolled grazing.

Greece

For this very rugged and broken-up country, it is not easy to mark climatic limits on a 1 : 25,000,000 scale map.

Climates

According to Mariolopoulos *et al.* (1964), 'one notes that Greece is less arid than eastern Spain or Asia Minor, but more arid than Peninsular Italy'. The regions mapped as semi-arid correspond closely to previous classifications (Mistardis, 1960, 1962), especially the eastern Peloponnesus, part of Euboea, the Thracian plain, the island of Crete, the Cyclades, etc.

The rainfall regime is practically the same as in the arid regions of Sardinia, Sicily and southern Italy (pronounced summer drought). The stations in the Cyclades are generally the driest (seven dry months at Thira and Naxos, see Fig. 1). A few stations have cool winters (5° C in January at Salonica and at Larissa in Thessaly, see Fig. 1), and hot summers (about 27° C in July at these same stations).

Natural vegetation and land use

In the Cyclades, Athens region and eastern Peloponnesus, degraded vegetation of evergreen Mediterranean oaks with *Pistacia, Phillyrea, Olea, Myrtus,* etc., is found. On Crete and some islands of the Aegean Sea, the *Pinus halepensis* and *P. brutia* forests have been badly degraded (less than 3 per cent of the area). The degraded vegetation types or 'phryganas' include a common *Rhamnaceae, Paliurus spina-christi,* and *Poterium spinorum*. The Thracian plain, the Salonica region and Thessaly belong to another group, that of 'sub-Mediterranean' oaks (*Quercus pubescens* and *Q. conferta*).

Table 1 summarizes data on land use, using agricultural statistics for 1950 (Yassoglou *et al.,* 1964).

TABLE 1. Land use in semi-arid provinces in Greece (Yassoglou *et al.,* 1964)

	Attica	Argolis Corinth	Cyclades	South-east Crete
Land use (percentage)				
Agriculture	39.5	32.6	20.0	30.0
Forest	18.4	20.0	0.5	2.0
Pasture	20.6	37.0	63.0	
Crops (percentage of cultivated land)				
Annual crops	64.0	68.0	66.0	54.0
Vineyards	18.0	11.0	12.0	13.0
Olive	4.0	7.0	3.0	21.0
Citrus	0.5	1.5	1.0	0.5

Wheat and barley are the commonest annual crops, grown on 50 to 60 per cent of the cultivated area. In Greece, as in many dry Mediterranean countries, there is a pressing need to reduce the number of goats and speed up reafforestation.

Maghrib countries and Libyan Arab Jamahiriya

Climates

These countries have been unequally studied. In the comprehensive bibliography on arid zones of Ionesco (1965), which includes 542 references, it is evident that Morocco is the most studied country, thanks especially to the works of Emberger and Sauvage. Ionesco (1965) has published about ten 1 : 4,000,000 scale maps on which the different aridity indices (Koppen, Emberger, Gaussen, Meigs, Thornthwaite, etc.) are applied to Morocco. The relative utility of each climatic index has been raised recently by numerous authors, especially by Metro (1970) who compares several indices (Gaussen's xerothermic index, Thornthwaite's global index, Emberger's pluviometric index).

At the coastal stations, the rainfall regime is characterized by a single rainfall maximum in winter, and almost total summer drought (Algiers and Tunis, Fig. 3). Inland and in the mountains, there is a widespread regime with two rainy seasons, in spring and autumn (Marrakech, Fig. 3). It should be noted that there is a sharp decrease in rainfall from north to south on the western coast (Tangier, 820 mm; Rabat, 500 mm; Casablanca, 400 mm (see Fig. 3), Essaouira, 330 mm). Algiers receives 650 mm of rain, and Tunis 420 mm. The mountains considerably affect rainfall distribution. In Algeria and Tunisia, the proximity of uplands to the sea allows a Saharan influence to reach a high latitude.

The influence of the desert climate is felt even further north along the Libyan coasts and around Alexandria, where the mean length of the dry period is between eight months (Benghazi) and ten or eleven months (Tripoli, Ajedabya). On the other hand, northern Morocco is sheltered by the Atlas mountains, and is hardly affected by this influence. At the great majority of stations, the temperatures fall in the group with cool winters (0° to 10° C). At low altitude the summers are hot (20° to 30° C), while on the uplands they are 'temperate' (10° to 20° C). Summers are hot or very hot in southern and eastern Tunisia.

Natural vegetation and land use

A relationship can easily be established between the climatic map of arid regions and the vegetation stages defined by Emberger (1951) and Ionesco (1965).

The dry northern parts of North Africa correspond mainly to the wooded 'semi-arid Mediterranean stage'. Wooded pseudo-steppes or more or less degraded woodlands of *Tetraclinis articulata* or *Juniperus phoenicea* are found there.

The hyper-arid southern areas correspond to the non-wooded 'Saharan stage' with some woody-vegetation steppes and salt steppes.

The relationship between soil and vegetation is most obvious in the driest areas. Emberger (1951) has established some relationships, which are shown in Table 2. Unesco-FAO (1970) gives details of the dominant plant formations in the various desert regions mapped.

Between the hyper-arid Saharan zone, and the less arid regions in the north, is the non-wooded 'arid Mediterranean stage' which includes some typical steppes distinguished by the dominant plant

TABLE 2. Relationships between soil and vegetation (after Emberger, 1951)

Climax groups of the Saharan Mediterranean Stage (plant associations with)	Soil conditions of the stage
Nerium oleander	Oases, permanent or subpermanent water points
Aristida pungens	Sands (nebkas), dunes
Nanophanerophytes (*Ephedra alata, Anabasis*), hemicryptophytes (*Andropogon laniger, Aristida obtusa*) and geophytes; numerous therophytes	Clay and gravel steppe (reg)
Rhus oxyacantha and *Warionia*	Rocky steppe (hamada)
Pistacia atlantica-Ziziphus	Depressions
Tamarix articulata	Underground water course
Arthrocnemum sp. and *Halipedes* sp.	Chotts, salty ground

species: woody plant steppes with *Artemisia herba-alba*, with *Ziziphus lotus*, or with *Arthrophyton scoparium*; the so-called 'salt' steppes (Sauvage and Ionesco, 1962) with *Frankenia* and *Suaeda fruticosa*; grassy steppes with *Stipa tenacissima*; shrubby steppes with *Argania spinosa*, with *Pistacia atlantica,* and with *Acacia raddiana*. Table 3, from Emberger (1951), shows the detailed relationships between climax plant associations and vegetation stages in North Africa.

The land use of the different Maghrib countries is quite well known as a result of the detailed cartographic studies of Théron and Vindt (1955), Barry *et al.* (1963, 1973) and Quézel (1964, 1968). According to Depois (1964), if the hyper-arid Sahara is excluded, forests occupy 16 per cent of Morocco, 12.5 per cent of Algeria and 9.5 per cent of Tunisia. Tree crops (olives, figs, apricots, almonds, walnuts) are particularly developed on the uplands. Cereals, winter crops especially, are essentially grown in the lowlands, except for barley and rye which are found in the mountains and high plateaux.

TABLE 3. Relationships between climax plant associations and vegetation stages in North Africa (after Emberger, 1951)[1]

Climax association	Stages							
	Saharan	Arid		Semi-arid			Sub-humid	Humid
		Mild	Cold	Mild	Average	Cold		
Juniperus thurifera						+		
Juniperus phoenicea				+	±			
Tetraclinis articulata				+				
Cupressus sempervirens					+			
Pinus halepensis				+	±			
Quercus ilex					+	+	±	±
Quercus suber				+	+		±	±
Ziziphus lotus	±	±	+	+	+		+	
Acacia gummifera		±		+				
Acacia tortilis	±	±						
Oleo-lentiscetum					±		+	+
Argania spinosa		±		+				
Pistacia atlantica		±		+				
Stipa tenacissima			+					
Artemisia herba-alba			+					
Continental associations with *Halipedes* sp.	+	+	+	+	+	+		

1. ± = preferred stage.

Countries of the Near and Middle East[1]

Turkey

Two very characteristic climatic types can be distinguished. The first type covers the continental and mountainous regions of central and eastern Anatolia, with the stations of Ankara (Fig. 3), Kayseri and Sivas (Fig. 3). Rainfall is low (generally < 400 mm), spread over two not very rainy seasons in spring and autumn. The dry season lasts about 5 months (June to October), winters are cold (a January average of less than 0° C), and summers hot (a July-August average from 20° to 25° C). The eastern half of the country is mainly made up of high mountains, and agriculture is only possible in the lower areas (less than 10 per cent of total). This is a region of extensive sheep and cattle pastoralism. However, cereal crops (mainly wheat and barley) occupy large areas of the lower plateaux of central Anatolia. In the western continental regions *Pinus nigra* (*P. pallasiana*) formations are common, as well as degraded stands of oaks (*Quercus pubescens*, *Q. infectoria*, *Q. libani*). Large areas are covered with high steppes of *Astragalus* and *Acantholimon*.

The second type covers the coastal regions, and corresponds essentially to western and southern Turkey, to Cyprus and Bulgaria (stations at Balikesir, Istanbul (Fig. 3), Izmir, Canakkale). Between 500 and 700 mm of rainfall is concentrated in winter (December to February). The average length of the dry period is three or four months. Winters are cool (5° to 10° C), and summers hot (20° to 30° C). Irrigation is often possible in these dry areas on both the Mediterranean and Aegean coasts. In consequence, agriculture is extremely diversified. So-called 'industrial' crops (tobacco, cotton, poppy) are chiefly confined to the valleys opening on to the Aegean. However, the total surface area currently under irrigation in Turkey is only about 8 per cent and this is strictly limited to the coastal fringe (Aegean Sea), the Konya region and especially the Cukurca region, the richest in the country (cotton, citrus fruits, early vegetables). The natural vegetation of the coastal

1. This region has been mapped by Unesco-FAO (1970); there is a vegetation analysis by Zohary (1973); see also Kaul (1970).

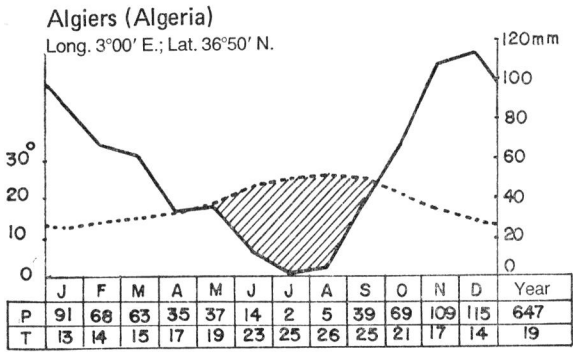
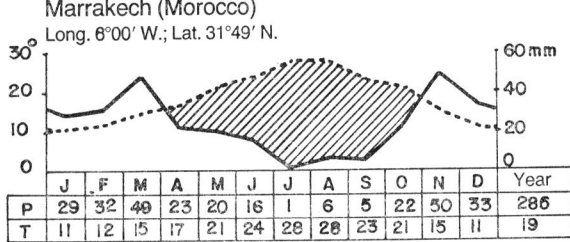
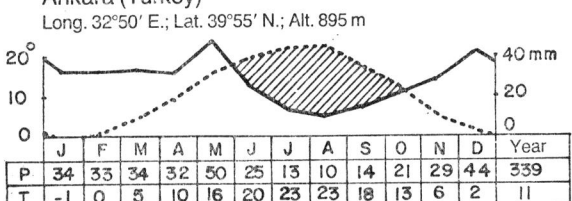
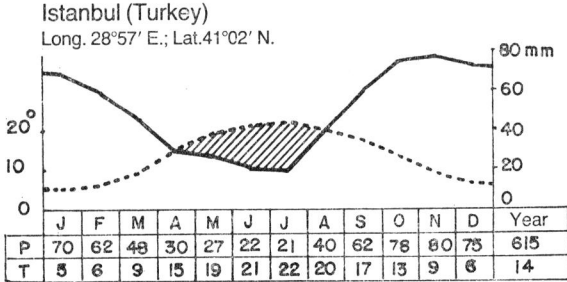
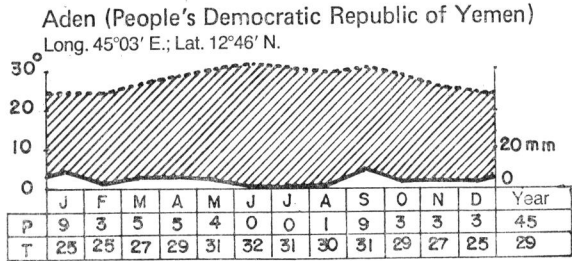
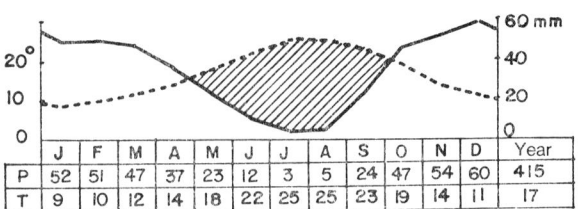
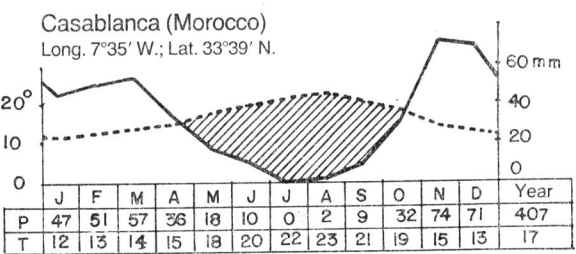
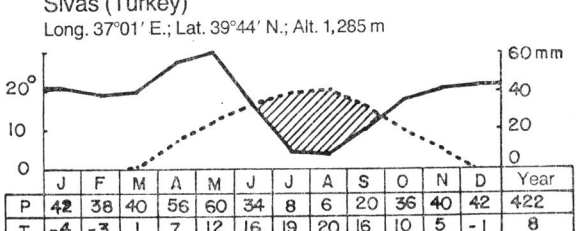
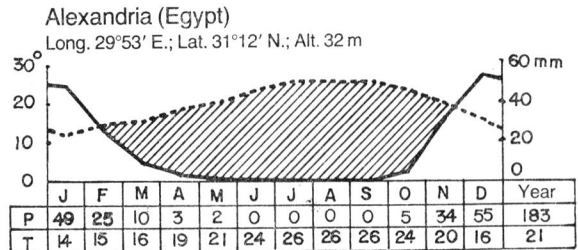
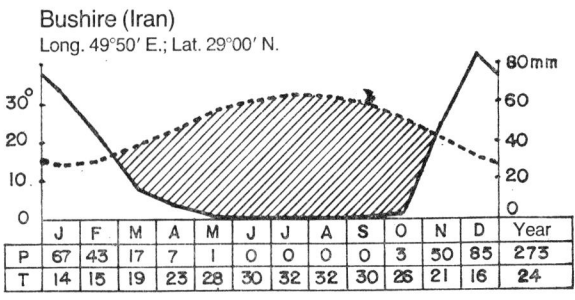

FIG. 3. Schematic representation of the climate at several stations in Algeria, Egypt, Iran, Iraq, Morocca, Syrian Arab Republic, Tunisia, Turkey, and Democratic Yemen.

Jask (Iran)
Long. 57°45′ E.; Lat. 25°45′ N.

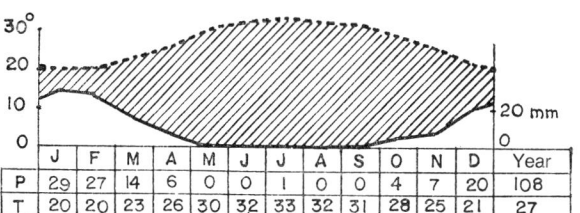

Isfahan (Iran)
Long. 51°41′ E.; Lat. 32°41′ N.; Alt. 1,650 m

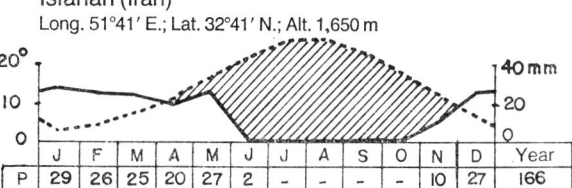

Sultanabad (Iran)
Long. 49°42′ E.; Lat. 34°05′ N.; Alt. 2,000 m

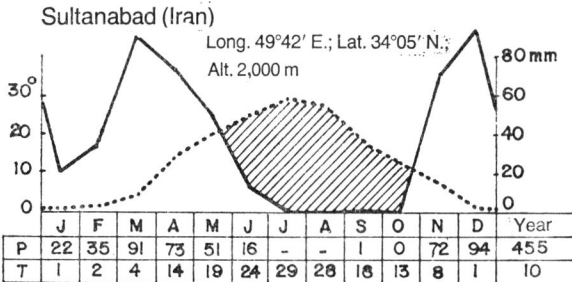

Deir-es-Zor (Syria)
Long. 40°11′ E.; Lat. 35°20′ N.; Alt. 200 m

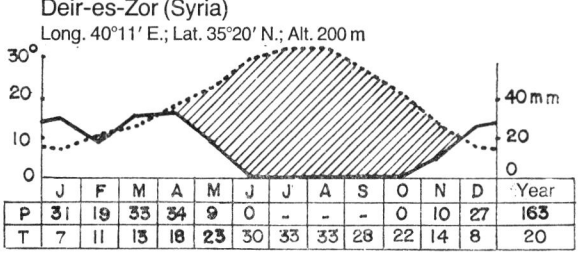

Mosul (Iraq)
Long. 43°05′ E.; Lat. 36°20′ N.; Alt. 223 m

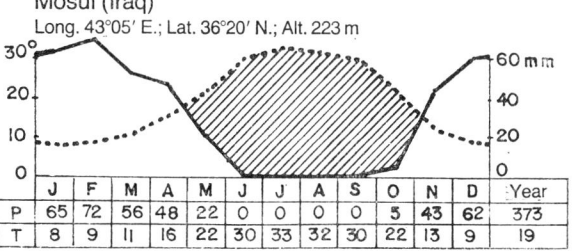

Tehran (Iran)
Long. 51°26′ E.; Lat. 35°40′ N.; Alt. 1,130 m

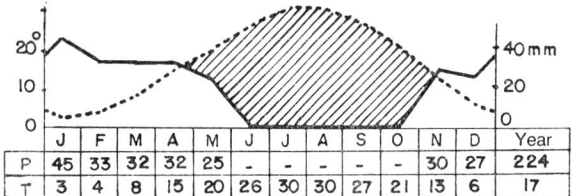

Hamadan (Iran)
Long. 48°36′ E.; Lat. 34°46′ N.; Alt. 2,300 m

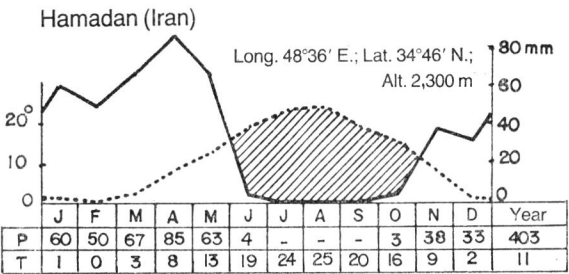

Palmyra (Syria)
Long. 38°15′ E.; Lat. 34°36′ N.

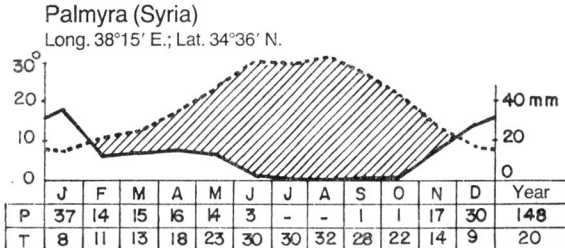

Rutba (Iraq)
Long. 40°18′ E.; Lat. 33°03′ N.; Alt. 615 m

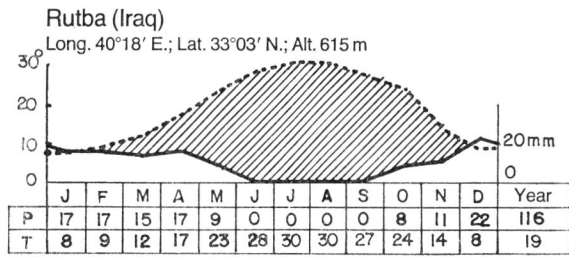

Baghdad (Iraq)
Long. 44°22′ E.; Lat. 33°20′ N.

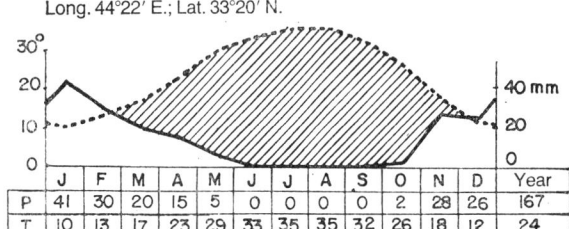

FIG. 3 *(cont'd)*

regions is often steppe-like: the Moldavian steppe (close to those of Crimea) has *Stipa, Artemisia, Astragalus* and *Festuca*; in Cyprus, a degraded maquis is to be found with lentisk, oaks (*Q. infectoria* and *Q. calliprinos*) and carob-trees. The vegetation of eastern Anatolia is typically Mediterranean.

More than 80 per cent of the Turkish population live from agriculture, and 90 per cent of the country's exports are of agricultural origin, but only 25 per cent of the country is cultivated because of physical limitations (climates, relief, steep slopes). See Erinc and Tundcdilek (1952), and the Turkish Atlas (Tanoglu, 1961).

Egypt

Egypt has an overall climate that is almost uniformly hyper-arid. Oekekoven (1970) defines it as follows: 'Alexandria [Fig. 3], the wettest part, receives only 184 mm of rain and most of the south has only 75 mm or less. In many districts, rain may fall in quantity only once in 2 or 3 years.' Summers are hot (20° to 30° C in July-August) and winters temperate (10° to 20° C). Sand-storms, light frosts, and morning fog are factors to be taken into account in a detailed climatic study. On the whole, Egypt is thus a very dry country. Only 5 per cent of the land is cultivated, limited to the alluvial soils of the Nile valley.

Despite the climatic aridity, the Egyptian flora is made up of several hundreds of species (Kassas, 1952, 1953, 1962; Tackholm 1954, Zohary, 1944), mostly ephemerals with bulbs or rhizomes (*Tribulus, Morettia*). In the low-lying areas, the scrub contains *Acacia futilis, A. flava, A. raddiana, Crotalaria aegyptiaca*, etc.

Reafforestation is of major importance. Until now, however, very small areas have been planted, using only a few species of *Eucalyptus* (*E. camaldulensis, E. citriodora, E. gomphocephala*), of *Casuarina* (*C. equisetifolia, C. glauca, C. cunninghamiana*) and also *Albizzia lebbek, Tamarix articulata, Dalbergia sissoo*, etc.

Saudi Arabia[1] and Democratic Yemen

This region contains three hyper-arid areas: the Great Nafud, the Rub Al-Khali, the Red Sea north of Jiddah and the coast north of Aden (Fig. 3). Natural vegetation is sparse in the three areas. Sporadic species such as *Calligonum comosum, Haloxylon salicornicum, Ephedra alata* and *Artemisia monosperma* are quite characteristic. On the coast north of Jiddah there is in places a very open thorn bush steppe (mostly *Acacia tortilis*).

The bioclimates of the People's Democratic Republic of Yemen and the Red Sea coast differ from the preceding because of high winter temperatures (25° C at Aden in January); summers are very hot (32° C in June). In spite of appearances the People's Democratic Republic of Yemen is on the whole quite suitable for agriculture. A clever system of terrace cultivation and irrigation allows a rational use of upland water resources and a fairly effective protection against erosion, especially on the western slope. The dry coastal plain is well cultivated, with irrigation from small seasonal or permanent water courses. This so-called Tihamah plain has few groundwater resources, but they are plentiful in the uplands (in the Sarnnoa, Mabar, Dhamar regions).

In the Hodeida area, mangroves (*Avicennia officinalis, Bruguiera gymnorhiza*) are among the characteristic species of the natural vegetation. The commonest shrubs inland as far as the foothills are *Salvadora persica, Cadaba rotundifolia, Tamarix articulata, Acacia flava* (= *A. ehrenbergiana*). On the slopes, low open forests of *Acacia nubica, A. hamulosa, A. verugera, A. spirocarpa* are note uncommon in the People's Democratic Republic of Yemen where a tropical flora of *Grewia, Commiphora, Dodonea* and *Ficus* is quite common.

Iran

Iran has the greatest climatic and phytogeographic diversity in this part of the world because of its size and important mountain ranges.

In addition to the hyper-arid Lut region studied by Dresch (1968), three other major regions should be distinguished. First is the coastal region of the Persian Gulf (Bushire, Fig. 3) and of the Gulf of Oman (Jask, Fig. 3), arid, with temperate winters (10° to 20° C) and with very hot summers. The vegetation consists of a spiny shrub and grass pseudo-steppe with almost tropical plants: *Capparis, Acacia, Calligonum, Ziziphus, Calotropis, Salvadora, Cenchrus ciliaris, Hyparrhenia hirtus*, etc.

Secondly, the central Iranian plateau, illustrated by the stations at Tehran (1,130 m, Fig. 3), Isfahan (1,650 m, Fig. 3), and Kerman (1,900 m), is very dry (230 mm, 165 mm and 170 mm respectively). There is no rain from June to October-November; rain falls from January to March or April. The

---

1. See Good (1954) for the vegetation of the islands of the Persian Gulf.

Great Kavir to the east of Tehran is drier: mainly cool winters (0° to 5° C), generally hot or very hot summers (26° to 31° C in July). The most typical vegetation is a plateau steppe with *Artemisia marina, Stipa lagascal,* with occasional tree or shrub forms with *Pistacia terebinthus* and *P. khinjuk, Amygdalus scoparia, A. horrida, Juniperus excelsa.* Mean annual rainfall in the Lut is put at less than 20 mm. Plants and animals are very rare in the central parts.

Third, in the mountains that stretch almost from Lake Ourmia in the north-west to Hashtom (Kerman region) in the south-east, temperatures are lower (cold winters, with the mean temperature for December, January and February at or below 0° C). Summers remain hot (most often 25° to 30° C in July-August). The increase in rainfall (405 mm at Hamadan and 460 mm at Sultanabad, see Fig. 3) markedly reduces the number of dry months to 5 (from June to October). The vegetation is generally made up of various xerophilous oaks with *Quercus persica* dominant, *Q. lusitanica* and *Q. infectoria*, to which *Juniperus excelsa,* etc., is added on the southern slopes of the Elburz.

In eastern Iran, and also further east in Baluchistan (Ladgasht region) and Afghanistan (Zirreh), is found a low vegetation of salt-tolerant plants (*Halocnemum, Salicornia, Salsola, Aeluropus repens*).

Syrian Arab Republic, Iraq, Kuwait

Where annual rainfall is around 100 to 150 mm or less, the vegetation is practically that of desert regions. This is the case of almost all of Kuwait where there is a severe lack of fresh water, even for human consumption. Desalinization of sea water meets domestic requirements, but irrigation is only possible using well-water which is almost always brackish; some vegetables, lucerne, barley and date palms are able to grow with this water. The recent discovery of fresh water springs in the Rawdahtain area 60 km north of the town of Kuwait, gives hope that the town can be supplied from there.

Where annual rainfall is from 150 to 300 mm, the vegetation is usually a 'dry steppe', with Mediterranean influences, with *Artemisia herba-alba, Stipa tortilis, Poa sinaica, Anabasis* and *Thymelaea hirsuta.* For Zohary (1973) these regions correspond to the 'Mesopotamian steppes with *Artemisia herba-alba'.* These vegetation types are very common north of the Nafud in Syria (Palmyra and Deir-es-Zor, see Fig. 3) and in Iraq (Rutba, see Fig. 3). The dry season is very long (nine to ten months), winters cool (5° to 10° C), and summers very hot (> 30° C).

Where annual rainfall is from 300 to 600 mm, the vegetation is a 'humid steppe', in which appear small woody plants (*Prosopis stephaniana, Ziziphus lotus*) and even almonds (*Amygdalus sparteoides*). The station at Mosul (see Fig. 3) in Iraq illustrates these climates, in which temperatures vary little from the previous stations, but where winter and spring rains are effective. The dry season is reduced to six months in summer (May to October).

On the uplands, oaks are dominant once the rainfall exceeds 600 mm (*Quercus aegilops, Q. infectoria* at low altitude, *Q. libani* above 1,300 m). The presence of *Pinus brutia* stands 60 km north of Mosul should also be noted.[1]

This general overlap between climate and dominant vegetation is altered by soil and management. Thus the trees and shrubs like *Populus euphratica, Salix* and *Tamarix* sp., the ecology of which is closely linked to soil moisture, are independent of regional climatic variations. The lower valley of the Tigris and the Euphrates, which is very dry (Baghdad, Fig. 3) but has considerable soil water reserves, has been transformed into a huge oasis. Iraq is the world's largest date producer (80 per cent of world production). Only 3 per cent of the total surface area of Iraq is cultivated, and extensive agriculture gives poor yields in spite of general fallowing. The main crop is barley, followed by wheat.

Lebanon, Israel, Jordan

These three countries, especially Israel and the Lebanon, are among the best studied dry countries. There are very many works on their phytogeography climate. Chouchani (1972) provides a comprehensive bibliography and an up-to-date synthesis for the Lebanon. The best known works on Israel are those of Karschon (1954, 1955, 1961a, 1961b, 1964a, 1964b) and of Zohary (1962). A good summary of the various climatic types and land use in Jordan is given by Oedekoven (1970).

Zohary (1973) attaches the overall vegetation of these regions to the 'Mediterranean woodland climax' which includes: (a) *Quercetea calliprini* (including *Pinus brutia* forests) and *Sarcopoterietalia spinosae*; (b) *Quercetea cerris oromediterranea orientalia* (including *Pinus nigra* forest). This designation is virtually equivalent to the 'formations of the western Mediterranean evergreen oak stage' of Unesco-FAO (1970). Climax formations at low altitude have

---

1. See the botanical studies by Guest (1932, 1933), Gillet (1948) and by Guest and Al-Rawi (1960).

disappeared or are very degraded. In Jordan, only 0.4 per cent of the land is forested, with *Quercus calliprinos* generally making up 75 per cent of the forest stands.

In Israel, 75 per cent of the land is under cultivation, of which 40 per cent is irrigated (Amiran 1964).

Afghanistan

This is a relatively little studied country. In 1973, the *Bibliography on Plant Ecology in Afghanistan* contained only eighty references. Three main climatic regions can be distinguished.

The hot arid region occupies the lowland regions of the south and south-west and, in the north, a narrow strip along the U.S.S.R. border. One of the driest parts of the country is in fact found in the south-west near the Iran-Afghanistan border (Farah and Chakhansur). In places mean annual rainfall can be less than 100 mm. At Kandahar, at an altitude of 1,000 m, rainfall is 145 mm/year, with a December-April maximum. The dominant vegetation type is a very open steppe with *Aristida plumosa, Arthrophyton persicum* and various *Calligonum*. In the north of the country, the Maimana and Kunduz stations are markedly wetter, with 370 and 350 mm/year respectively.

The cool, semi-arid region, containing Kabul (altitude 1,800 m, rainfall 350 mm), includes to the south and west the mountains of north Afghanistan. Rain falls from February to May. The natural vegetation is a mid-altitude steppe, sometimes with trees (*Pistacia, Juniperus, Amygdalus*), composed of small highly adapted plants of the genera *Cousinia* (*Compositae*) and *Acantholimon* (*Plumbaginaceae*).

The cold sub-humid upland region is in the Hindu-Kush, especially on the Afghan Pamir. At Lal, at an altitude of 2,800 m, 280 mm/year of rain falls with an April maximum, as at Kabul. The steppes are no longer wooded; the genera *Cousinia, Astragalus* (*Leguminosae*), and *Festuca* are well represented.

The unfavourable ecological conditions in Afghanistan are aggravated by a combination of two factors: it rains when it is cold (mainly February to April), which means that the water cannot be optimally used; and the summer is extremely dry. Thus the distance covered by transhumant pastoralists is considerable, from Baluchistan (Pakistan) up to the Afghan plateau.

Central Asia

The main arid regions of the U.S.S.R., China and Mongolia have been the object of relatively little known but detailed research (Gao, 1962; Ivanov and Vakulin, 1962; Kachkarov and Korovine, 1942; Leont'ev, 1962; Petrov, 1952, 1957, 1970a, 1970b).

Main desert regions of the U.S.S.R.

These cover very large areas north and east of the Caspian Sea almost as far as the Altai. They comprise in particular:
Turkmenistan, 87 per cent of the surface of which is covered by the Kara Kum sands, known as '*Haloxylon* deserts'.
Uzbekistan, south-east of the Aral Sea between the Amu-Daria and the Syr-Daria (Kyzyl Kum desert).
Kazakhstan, where a steppe vegetation of 'wormwood and grass' dominates, but which does have areas of desert sands comparable to those of the Turkmenistan, especially north of the Caspian Sea, east of the Aral Sea and south of Lake Balkash.
The deserts of the U.S.S.R. are on the whole considered as economically rich regions. There are considerable hydrocarbon reserves, for example in the Tashkent region and south-east of the Caspian Sea. Coal (in the Nukus and Termez areas), copper (at Lake Balkash) and other raw materials are not rare. These areas have long been known for sheep and horse raising.

Climates

Figure 4 shows a schematic representation of the climate at several stations in China, Mongolia and the U.S.S.R. Continentality is one of the main climatic characteristics of these deserts. For example, it is in these regions that the highest temperatures in the U.S.S.R. occur (Turkmenistan, July average 32° C). Winters are cold: — 26° C has been registered at Ashkhabad at latitude 38° N. (altitude 220 m), and the January average for Chelkar near the Aral Sea is about —15° C (July average 26° C).

Therefore, the dominant 'class' is that of cold winters and hot summers. Several representative

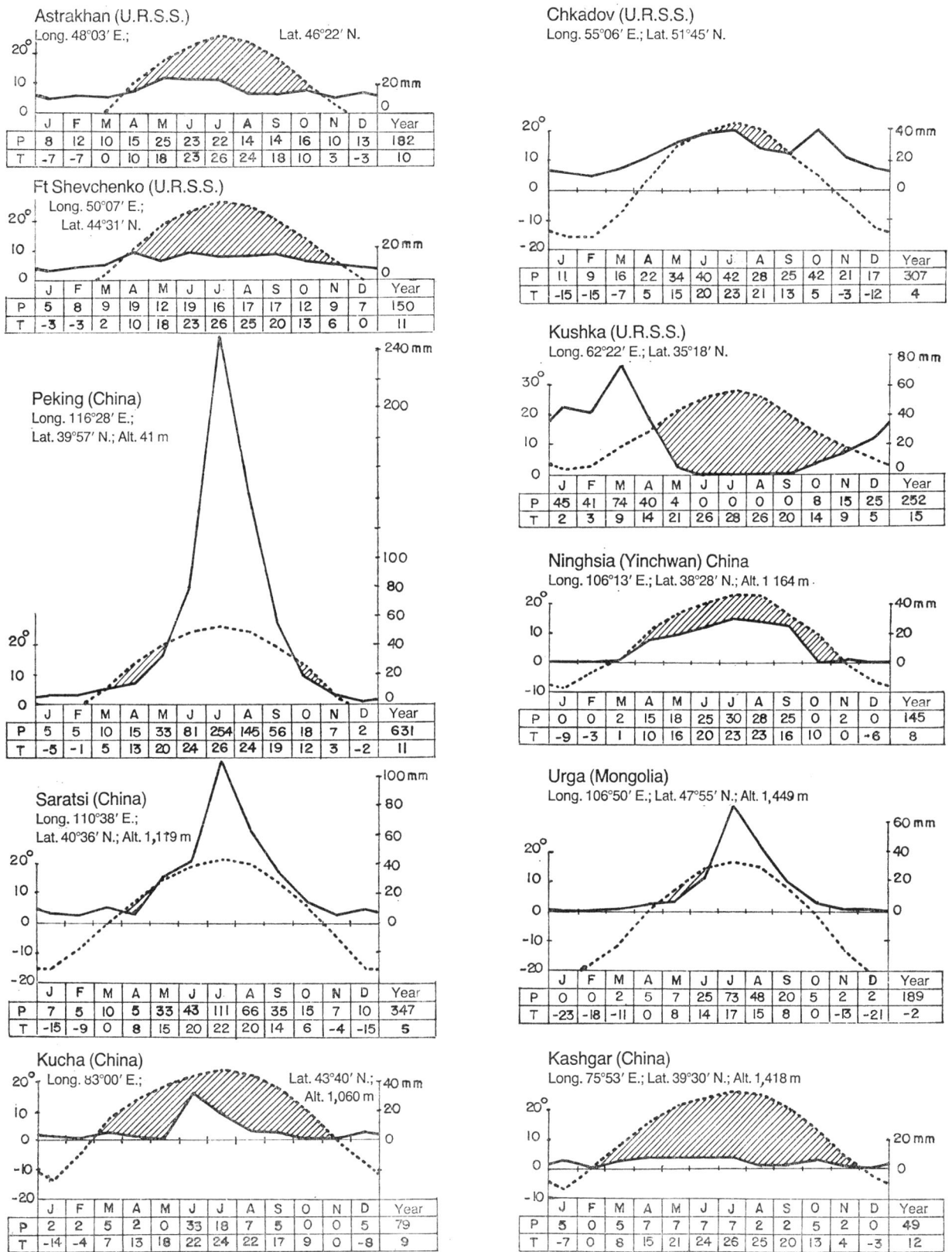

FIG. 4. Schematic representation of the climate at several stations in China, Mongolia and the U.S.S.R.

stations can be cited: Astrakhan (Fig. 4), Chkadov (Fig. 4), Turgai, Fort Shevchenko (Fig. 4). A 'temperature class' of smaller size, with cool winters (t_m about 3° C) and very hot summers, is found in southern Turkmenistan ($t_x > 30°$ C at Kazandjik, Ashkhabad, Repetek, Kushka, Fig. 4).[1] Finally, north of the Aral Sea and in the mountains of Kirghizistan east of Tashkent, narrow strips of dry land are characterized by cold winters ($t_m = -19°$ C at Omsk and Naryn) and temperate summers ($t_x = 18°$ to 19° C).

Dominant vegetation

There are two basic types of dominant vegetation: sandy deserts with saxaul in Turkmenistan and Uzbekistan; and *Artemisia* steppes in Kazakhstan. In general the plant species have great ecological plasticity. They are well adapted to drought, to winter cold, and for many, to high concentrations of salt at the soil surface. The forestry services have selected indigenous species able to constitute a woody plant cover to stabilize sand. Good results have been obtained and the present knowledge of plant ecology is due to their work. The genera *Calligonum* (Polygonaceae) includes more than ten species of small to tall shrubs which are naturally or artificially distributed in these regions. *Calligonum caput medusae*, *C. arborescens*, *C. eriopodum*, *C. turkestanicum* and *C. aphyllum* are the best known. *Haloxylon* (Chenopodiaceae) are probably the most characteristic shrubs of these deserts. Better known as 'saxaul', they are represented by two species, *H. aphyllum* (6 to 8 m) and *H. persicum* (4 to 5 m), widely planted to combat wind erosion. Two other woody Chenopods, *Salsola richteri* and *S. paletzkiana* have similar ecologies but these latter are more salt tolerant.

Numerous *Tamarix* are also planted in the Soviet deserts (*T. ramosissima*, *T. laxa*, *T. szovitsiana* and *T. bungei*). Among the other important naturally occurring woody plants there are also legumes (*Ammondendron conollyi*, *A. karelini*), sagebrushes (*Artemisia arenaria*), milk-vetches (*Astragalus paucijugus*, *A. unifoliolatus*), which are good fodder plants. Two psammophilous grasses are also widespread: *Aristida karelini* and *A. pennata*.

A fairly complete list of xerophytes and salt tolerant species currently used in the deserts of the U.S.S.R. for sand fixation and for rangeland improvement is found particularly in the works of Petrov (1950, 1952, 1957, 1970a).

Hydrology and land use

The natural drainage system is very unusual. Numerous streams rise in the southern uplands and therefore flow south to north, disappearing gradually in the deserts by infiltration or evaporation. Some, however, manage to cross hundreds of kilometres of desert and supply the inland seas. This is the case of the Amu-Daria and the Syr-Daria which flow into the Aral Sea, and of the Ili and Karatal which feed Lake Balkash. Fresh-water reserves are abundant beneath the beds of detritus of the pediments, although those of the alluvial plains almost always have a high salt concentration.

The Soviet deserts have undergone major transformations as a result of numerous development schemes. Although the Kazakhstan steppes are still above all pastoral regions, enormous areas which were for thousands of years considered sterile are today under irrigated cultivation. These total 4 million ha. The canal which crosses the Karakum desert, from Termez in the west to the Caspian Sea, is one of the most striking achievements.

Main desert regions of China and Mongolia

The regions discussed here are commonly known under the names of the deserts of Ordos (at the mouth of the Hwang), Alashan (southern Mongolia), Pleishan, Takla-Makan (southern Sinkiang), Gobi and Dzungaria.

Climates

There are two main temperature groups: (a) regions with cold winters ($t_m < 0°$ C) and temperate summers (10° C $< t_x <$ 20° C), including large areas of the southern uplands (Tibet) and of the high plateaux situated in northern Mongolia; regions with cold winters and hot summers (20° C $< t_x <$ 30° C), stretching from Sinkiang in the west to the Shantung peninsula. A hyper-arid enclave has been marked in Sinkiang. This is the Charchan region, which is the most arid, where the mean annual rainfall is less than 10 mm.

Petrov (1970b) characterizes the climates as follows:

The Central Asian deserts lie in the moderate zone. According to the classifications of Thornthwaite (1948) and Meigs (1952), these territories should be classified

---

1. t_x, mean temperature of the hottest month; t_m, mean temperature of the coldest month.

as hyper-arid, arid and semi-arid regions (SbO_3, AaO_3, AcO_3, EbO_2, EbO_3).

The climate of the Central Asian deserts is moderately cold with mean annual air temperatures varying from 2.3° C in the mountain deserts of Tsaidam to 11.6° C in the warmest regions of Kashgar. The mean July air temperatures vary from 16.7° C in Tsaidam to 27.3° C in Khami (the Gashun Gobi) depending on the altitude.

The mean annual precipitations do not exceed 100 mm except in the eastern steppe regions and foothill plains. The deserts of Kashgar, Tsaidam and Pleishan, where precipitation does not exceed 10 mm per year (Cherchen), are the most arid areas. In the eastern areas of Central Asia, rain is a result of the eastern Chinese monsoon and has minima in summer, in Dzungaria and western Kashgar; the comparatively regular distribution the year is due to the invasion of the humid Atlantic air. The evaporation reaches 3,500 mm at Hami.

As the diagrams for Peking, Ninghsia (= Yinchwan), Saratsi (in the Suiyan) and Urga (in Mongolia) in Figure 4 show, the great majority of stations in this part of the world have a simple rainfall maximum in summer. This pattern is not apparent in the hyper-arid regions (Kuska and Kashgar in Sinkiang, see Fig. 4).

Large areas of ancient lakes are now covered with 'solonchaks' (white saline soils) without vegetation. In addition, the major sand deserts are in the west (Kashgar, Dzungaria);[1] rocky deserts, the most difficult to develop, are mainly found in the central Gobi desert.

Dominant vegetation

This varies considerably according to the physical and chemical properties of the soil and soil moisture.

The only shrub formations with *Haloxylon ammodendron*, *Salsola* and *Nitraria* are found in depressions where the ground water is close to the surface. *Phragmites* and *Lasiagrotis* occur wherever the ground water comes to the surface. If the water is saline, a 'specialized' flora with *Tamarix*, *Nitraria siberica* and *Kalidium foliatum* is often dominant (Norlindh, 1949). Where rain is the only source of water, the vegetation becomes extremely poor with such characteristic psammophile species as: *Artemisia ordosica*, *A. sphaerocephala*, various *Caragana* and *Hedysarum*; and in stony deserts (Gobi): *Nitraria phaerocephale*, *Calligonum mongolicum*, *Amnopiptanthus mongolicus*, etc.

In these pastoral regions, problems of overgrazing and shifting sand are resolved by a choice of species which can both provide forage and stabilize dunes; according to Petrov (1970) these are *Hedysarum scoparium*, *H. mongolicum*, *Caragana korshinskii*, *C. microphylla*, *Calligonum zaidamense*, *Artemisia ordosica*, *A. halodendron*, *A. sphaerocephala*, *Astragalus melilothoides*, *Agriophyllum gobicum*. Norlindh (1949) has published interesting floristic commentaries and photographs of the plant formations of these regions. Closed basin drainage is particularly pronounced here. Only one important water course, the Hwang Ho, reaches the China Sea.

The huge upland desert of Tibet, for which hardly any continuous climatic readings are available, is a unique case. This is an area of about 900,000 km² at an altitude ranging between 4,000 and 5,000 m. The driest area lies in the centre, and probably receives less than 1,000 mm/year. In these regions the soil is frozen at least six months of the year.

Indian sub-continent

In India and Pakistan, the areas with arid or semi-arid climates are estimated at 600,000 and 204,000 km² respectively. Many works have been devoted to them. A good analysis of research undertaken in Rajasthan is found in Gupta and Prakash (1975). The Central Arid Zone Research Institute at Jodhpur (India) is concerned with all the problems of these regions.

There are semi-arid and sub-humid regions in India as for south as Sri Lanka. This is why it is necessary to refer to works covering the whole of the Indian sub-continent, notably those of Legris (1963), Spate and Learmonth (1967), Champion and Seth (1968), Mani (1974). The vegetation maps which give the best general ecological information are those collected in the *Carte Internationale du Tapis Végétal* (Gaussen et al., 1964). For bioclimates, Legris and Viard (1961), Labroue et al. (1965) have mapped a fairly precise distribution at about 1 : 2,500,000.

A simple presentation of the climates of the sub-continent makes it necessary to omit the humid or very humid uplands and western plains, as well as eastern India (Madhya Pradesh, Orissa, Bengal,

---

1. The famous Takla-Makan alone has nearly 350,000 km² of sands.

Bihar). The driest parts are found in the north-west, especially in Rajasthan and the Indus Valley in Pakistan. On the other hand, despite the name 'Thar desert', no hyper-arid region exists. Three major types can be distinguished.

Very hot, relatively dry regions

These very hot regions are found in peninsular India and Sri Lanka. They are mapped on the Deccan plateau, in the south of India (Tamil Nadu) and in small areas in the north and south-east of Sri Lanka (Mannar and Hambantota).

With the exception of the Ghat mountains, temperatures are quite uniform throughout these regions. They can be grouped in the class of hot winters ($t_m > 20°$ C) and very hot summers ($t_x > 30°$ C). A large part of the Deccan is semi-arid, the remainder sub-humid or humid. In peninsular India the dominant rainfall regime is of summer rain, although the rainfall maximum is sometimes delayed until September-October (Bellary, Fig. 5); in the south, particularly on the eastern coast, the maximum occurs as late as November-December (Blasco and Legris, 1973).

The natural vegetation of these regions has greatly suffered from considerable human pressures. Only degraded stages remain, in which weed species, unpalatable to livestock, are very common: *Dodonea viscosa, Cassia auriculata, Anisomeles malabarica,* various *Jatropha* and *Euphorbia,* etc. Savannahs are rare, although some dry deciduous forests remain. These contain *Albizzia amara, Tectona grandis,* various *Terminalia, Anogeissus latifolia,* etc. In some areas near the eastern coast there remain some tiny patches of a more or less evergreen thorny scrub with *Manilkara hexandra, Maytenus emarginata, Maba buxifolia, Hugonia mystax,* etc.

All these regions are highly cultivated. Irrigation using wells, tanks, and canals is very widely developed. The agrarian landscapes of the south are quite characteristic with a very common palm tree, *Borassus flabellifer,* and *Acacia leucophloea,* among the non-irrigated crops (millet, groundnuts). On the black basaltic soils of the Deccan, where millet and cotton predominate, the commonest trees are probably *Acacia arabica* and *Phoenix sylvestris.*

Hot, dry regions of the north-west

Drought increases in north India from east to west (see Fig. 5). New Delhi is in a sub-humid area; Bikaner and Barmer mark the transition between semi-arid and arid. Finally the stations of the Indus Valley in Pakistan are mostly in an arid area. Multan and Hyderabad receive 180 mm/year, Jacobabad less than 100 mm/year. All these regions have temperate winters ($10° < t_m < 20°$ C) and very hot summers (33° to 34° C in May and June in New Delhi, 35° to 37° C at Jacobabad). The most frequent rainfall regime in the east is a tropical one with a single summer maximum. However, in the west, in Pakistan, the stations at Quetta and Pasni have very dry summers. Meher-Homji (1963, 1974) has studied the variations in rainfall from year to year at one station.

The natural vegetation in these regions includes several characteristic physionomic and floristic types. In the western parts, scrub woodland and thorny scrub with *Anogeissus pendula, Acacia catechu* and *A. senegal* are still fairly widespread especially on the slopes of small hills. The companion shrub flora is rich in species: *Prosopis cineraria, Capparis decidua, Acacia nilotica* subsp. *indica, Ziziphus mauritiana,* etc. The scrub is generally degraded and the flora is characterized by the presence of *Balanites aegyptiaca, Salvadora persica, S. oleoides, Acacia nilotica, Ephedra foliata, Calotropis procera,* etc.

In the arid sandy regions of western Rajasthan, *Calligonum polygonoides* on dune slopes, *Haloxylon salicornicum* in the interdune spaces and various psammophytes (*Aerva, Cyperus arenarius*) are noteworthy. The dunes of the Bikaner region have been fixed by an introduced Acacia (*A. spirocarpa,* var. *tortilis*); in the Jaipur area, it has even been possible to fix the dunes and then cultivate them in terraces, giving quite good results for wheat.

Crops are widely grown in spite of the lack of water and are mainly millet (*Pennisetum typhoides*) and pulses (*Phaseolus aconitifolius*). In arid Rajasthan the average density is about 40 inhabitants/km^2.

A general idea of land use in arid India (percentage of the total area) is given by the figures of Kaul (1970): forest, 1; bare uncultivable land, 14; rangeland, 4; arable land but not cultivated, 21; cultivated land, 42 (of which 2 per cent is irrigated); miscellaneous, 18.

In the Kathiawar peninsula, south of Rajasthan, the vast flat, silty areas of the Great and Little Rann of Kutch are periodically invaded by the sea during the monsoon. The vegetation is very sparse or absent, except on the fringes where salt-tolerant grasses can grow: *Aeluropus lagopoides, Sporobolus* sp., etc. Blasco (1975) has provided a bibliography for these salty environments. The low-lying ground is flooded in the rainy season and salt-encrusted in the dry season. *Prosopis juliflora* is successfully planted there.

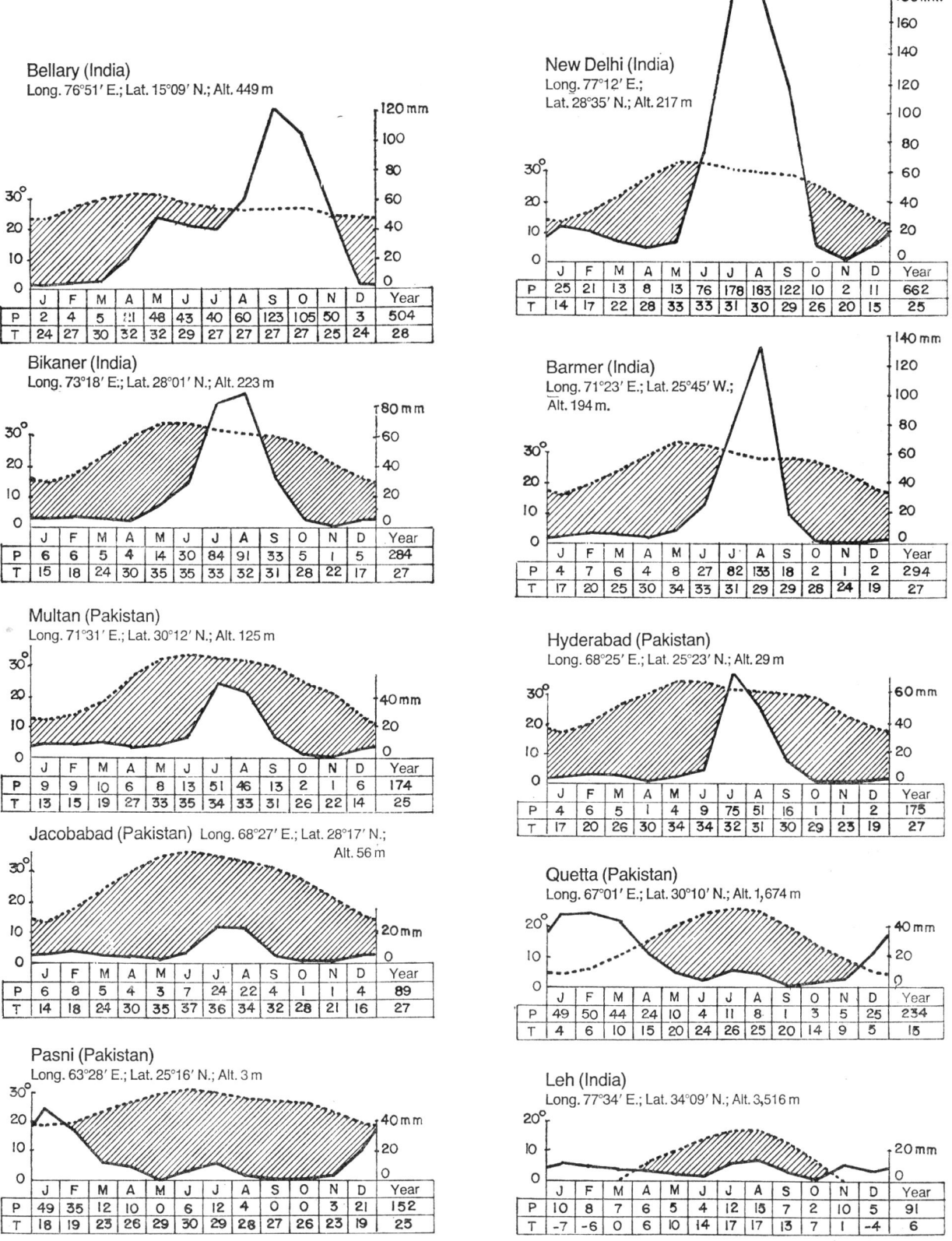

FIG. 5. Schematic representation of the climate at several stations in India and Pakistan.

Cold desert regions

These have been mapped in the high valleys of Kashmir, on the borders of Tibet. There is hardly any information on these areas. The works of Troll (1959, 1960, 1967) give some details on plant geography. The station at Leh (Fig. 5), at an altitude of 3,500 m, gives an idea of the aridity of the climate: winters are cold (—7° to —10° C in January), summers are temperate (a July-August mean of 17° to 18° C). These are regions of 'alpine steppes', grazed by yaks, with bushes of *Caragana pygmaea* (*Leguminosae*), *Ephedra nebrodensis, Hippophae* sp. (*Elaeagnaceae*), *Myricaria* sp. (*Tamaricaceae*), various species of *Salix, Populus, Juniperus,* etc. In Kargil and Ladakh, there are quite old plantations of willows and poplars covering about 150 ha.

Australia

At a continental or regional scale, the land resources, biogeography, ecology and land use of the arid and semi-arid lands of Australia are relatively well known. Many published works on these subjects at these scales are available (Burbidge, 1960; Keast *et al.*, 1959; Leeper, 1970; Moore, 1970; Stace *et al.*, 1968; Slatyer and Perry, 1969; Stephens, 1961; Wadham *et al.*, 1964). A great deal of other detailed information on the land resources of particular areas can be found in publications of the Commonwealth Scientific and Industrial Research Organization, particularly the CSIRO Soils and Land Use Series and the CSIRO Land Research Series. The *Atlas of Australian Resources* published by the Australian Department of National Development (1970) contains many maps with accompanying notes showing the spatial distributions of both physical and socio-economic features over the entire country.

Climates

The principal works on the regional pattern of climate are those of Gentilli (1953*a*, 1953*b*, 1971, 1972) and of the Bureau of Meteorology, which has published a concise but comprehensive general description of the climate of Australia (Bureau of Meteorology, 1974) as well as a series of climate surveys of particular areas.

A number of methods have been used to delimit climatic types in Australia, and a detailed review of these has been given by Gentilli (1972). The methods used range from simple use of average conditions of the elements, singly or in combination, to those attempting to integrate the principal elements, commonly by the use of single-value indices to give a quantitative assessment of precipitation effectiveness.

The classification of Köppen (1936) has been most widely used. Although this is especially useful for comparison at global scale, it fails to separate some areas having distinctive temporal and dynamic qualities in their climate. Such classifications as those of Emberger (1959) and Gentilli (1972) are more suitable. Both of these take cognizance of basic climatic controls operating in different areas as well as the intrinsic characteristics of climate revealed in the data for the major elements. The broad similarity in these two schemes is evident in Table 4. It is also seen that the divisions recognized by these workers are closely in accord with those shown on Unesco's *Map of the World Distribution of Arid Regions.*

Somewhat more than 80 per cent of the continental area of Australia is classified as arid or semi-arid (Meigs 1953), a larger proportion than for any other continent. Some appreciation of the vastness of this arid area can be had from the areas of the separately defined deserts within it: Great Sandy Desert (and Gibson Desert), about 600,000 km^2; Great Victoria Desert, 350,000 km^2; Simpson Desert, 800,000 km^2. Although Australia has the unwanted distinction of having the largest proportion of arid lands, it nowhere has the extreme conditions of aridity (hyper-arid climates) that occur in other continents. Indeed, as noted by Gentilli (1971) much of the area classified as arid is only marginally so, and a slight increase in rainfall would transform it into a semi-arid environment.

The lack of extreme dryness is clearly evident in the not uncommon and widespread occurrence of a relatively large percentage of ground cover by vegetation, and in the permanent pastoral land use found far into the interior of the continent. The basic causes of these less extreme conditions of aridity are the relatively small size of the island continent, the extensive surrounding oceanic regions which provide sources of maritime air that on occasion penetrate to the interior, the absence of very high orogra-

TABLE 4. Comparison of Emberger and Gentilli schemes for delimiting climate types in Australia

Emberger (1959)	Gentilli (1972)
1. Monsoon climate of extreme north (Arnhem Land and Cape York Peninsula)	Monsoonal wet/dry north (Ch. 4)
2. Semi-arid tropical climate (south of No. 1)	Sub-humid/dry inland north (Ch. 5)
3. Trade wind dominated climate of the north-east coast	Trade wind coast (Ch. 6)
4. Subtropical climate with influence of westerly winds (south-east coast: Bundaberg to Gabo Island)	Subtropical east coast (Ch. 7)
5. Cool subtropical climate (south-eastern Australia)	Cool moist climates of Victoria (Ch. 11)
6. Cold climates (Australian Alps)	Tablelands vertical topoclimates (Ch. 8)
7. Subtropical sub-humid climate	Sub-humid eastern region (Ch. 9)
8. Subtropical climate with dry summers (south-western Western Australia, Eyre Peninsula and Adelaide region)	Winter-wet south west (Ch. 12)
9. Cool climate with dry summers (Cape Northumberland region)	Humid and semi-arid climates of South Australia and Western Victoria (Ch. 14.)
10. Semi-arid subtropical climate (north of No. 8 and west of No. 7)	Eastern semi-arid lowlands (Ch. 10) and winter-moist semi-arid (Ch. 13)

phic barriers across the path of prevailing winds, and the absence of such very cold ocean currents offshore along the western subtropical coast as are found with other continents.

A characteristic feature of the Australian arid zone is the broad transition from arid to semi-arid and then to sub-humid conditions outward from a large central core area of aridity. Nowhere do these changes occur with steep moisture gradients, a fact that reflects not only the general insignificance of orographic control of rainfall, but also the spatial variability of those occasional influxes of moist air originating over surrounding maritime source regions.

To the north-west and south of the central arid core area, there is little diminution of aridity with approach towards the sea. Along the Indian Ocean coast between Broome and Carnarvon, rainfall is usually precluded by a prevailing subsidence and by offshore winds that have been dried by a long trajectory over the dry interior. However, heavy rainfalls do occur over this area in the north-west at times in association with tropical cyclones originating off north-western Australia. Along the southern coastline at the head of the Great Australian Bight, the arid Nullarbor Plain spans about ten degrees of longitude with only a very narrow strip of semi-arid land immediately along the coast. This is because eastward-moving winter frontal systems which pass over this area do not normally produce significant rainfall here where the prevailing air masses are considerably drier than those usually found over the south-west of Western Australia and over the southern parts of South Australia and western Victoria.

The seasonal incidence of rainfall is an important aspect of the arid environments of Australia because of its ecological and land use consequences. The degree and form of rainfall seasonality over the continent has been studied by Fitzpatrick (1964) using harmonic analysis applied to mean monthly rainfall. Over all of Australia north of the Tropic of Capricorn, there is strong summer concentration of rainfall with a well-defined dry cool season. From about the Tropic line southward, the winter component of rainfall increases, particularly over the eastern half of the continent. However, winter rainfall remains less than that in summer except along the southern coast, western Victoria and the south-west of Western Australia. Only in these areas does the strong winter concentration of rainfall and summer drought conditions typical of the Mediterranean lands occur, and these features do not generally extend far northward into the semi-arid and arid areas of southern Australia. In describing the rainfall of the southern part of inland Australia, Gardner (1959) writes that 'on the average the wettest month lies between March and July, but it must be stressed this rainfall is not of marked periodicity'. It warrants attention, however, that although winter rains are not particularly conspicuous in southern interior areas, their ecological significance is considerably enhanced due to comparatively low rates of potential evapotranspiration from May to August inclusive. The marked southward increase in the effectiveness of winter rainfall for promoting plant growth is clearly evident in a study which simulates soil water changes beneath a typical xerophytic plant community over a long run of years in interior Australia (Fitzpatrick et al., 1967).

Climatic regime of the arid zone

This vast area spans the Tropic of Capricorn and includes parts of all the states on the Australian mainland. The northern and eastern boundary of the area is generally from 500 to 800 km from the coastline if the northern peninsular extremities of the Northern Territory and Queensland are ignored. In the south-west of Western Australia, the limit of this area follows approximately a line from Carnarvon on the Indian Ocean coast to a point approximately 200 km east of Esperance on the western side of the Great Australian Bight.

This area is distinguished not only by the low ratio P/ETP (0.03 to 0.20 as shown on the map), but also by the erratic incidence of rainfall. There is clearly a trend from summer rain/winter drought conditions typical of the north to increasingly effective winter rainfall southward over the continent. However, in this arid area rainfall is so erratic that it is to some degree misleading to attach much significance to either mean annual rainfall or seasonal regimes of rainfall based on mean monthly data.

The driest part of this area is found in the Lake Eyre Basin where mean annual rainfall is as low as 125 mm, and where periods of several years can occur without significant rainfall for plant growth. Throughout a large part of this area, the average number of days with recorded rainfall is less than twenty-five and the average variability of annual rainfall is in excess of 30 per cent. Somewhat lesser variability of rainfall (20 to 30 per cent) occurs along the southern margin where winter rainfall contributes a larger proportion of the total than in the central and northern parts.

In its temperature characteristics, the arid area has mild winters (t_m from 10° to 20° C), and summers ranging from very warm in the northern sector ($t_x > 30°$ C) to warm in the southern sector (t_x from 20° to 30° C). The warmest summer conditions occur in the north-west in the vicinity of Onslow, Western Australia, where the mean daily maximum temperatures in January exceed 40° C. Mean daily minimum temperatures in July are mostly in the range of 5° to 10° C. Along the northern limit of this area frosts are unknown, but across the southern sector there is an average length of frost period of from 50 to 100 days (Foley, 1945).

Characteristic rainfall and temperature relationships for different parts of the defined arid area are shown in Figure 6 (Alice Springs, Birdsville, Meekatharra, Tarcoola). The data for Alice Springs and Birdsville illustrate the conditions over much of the central and eastern parts of the arid area, where there is quite strong concentration of rain in summer months and relatively light rainfall in winter. The data for Meekatharra in Western Australia illustrate the case of rainfall of about equal amounts occurring in the summer and winter, and that for Tarcoola in South Australia shows generally low rainfall throughout the year. The southward decrease in summer rainfall is clearly seen in the comparison of graphs for Alice Springs (23° 49' S.) and Tarcoola (30° 43' S.).

Soils and vegetation in the arid zone

Several distinctive physiographic desert types occur within this area (Mabbutt, 1969). The most widespread is the sand desert, consisting either of sand plains or dunes, most of which are in the form of long parallel ridges between 10 and 30 m high and at intervals of about 100 to 150 m. Also widespread are mountain and piedmont deserts, and an extensive granite shield with erosional remnants of lateritic capping makes up a large part of the arid area of Western Australia. Smaller areas of stony desert, desert clay plains, and riverine desert also occur.

The soils of this arid area are varied and strongly reflect the character of parent materials in many instances. A concise description at continental scale is given by Hubble (1970), and the nomenclature given here is taken from that account. Most extensive are siliceous or earthy sands, particularly in Western Australia, South Australia, and the Northern Territory. Red earths, red-brown earths and desert loams (with marked texture contrast between surface and subsoil), and grey, brown and red clays are also widespread throughout the area in all states. Shallow sandy soils (lithosols) or shallow loams are found in a widely scattered pattern, particularly in central Western Australia and the Northern Territory. Across the southern sector from Western Australia through southern South Australia and western New South Wales are extensive areas of solonized brown soils, grey-brown and red calcareous soils, and calcareous red earths. In general, the soils of the arid area of Australia are notably poor in phosphorous and nitrogen, and in many instances there are quite close associations between soil type and vegetation.

The vegetation formations and grazing land types of this area have been described by Perry (1970). Although there are many local variations, four major types can be identified: arid hummock grassland, *Acacia* shrubland, shrub steppe and arid tussock grassland.

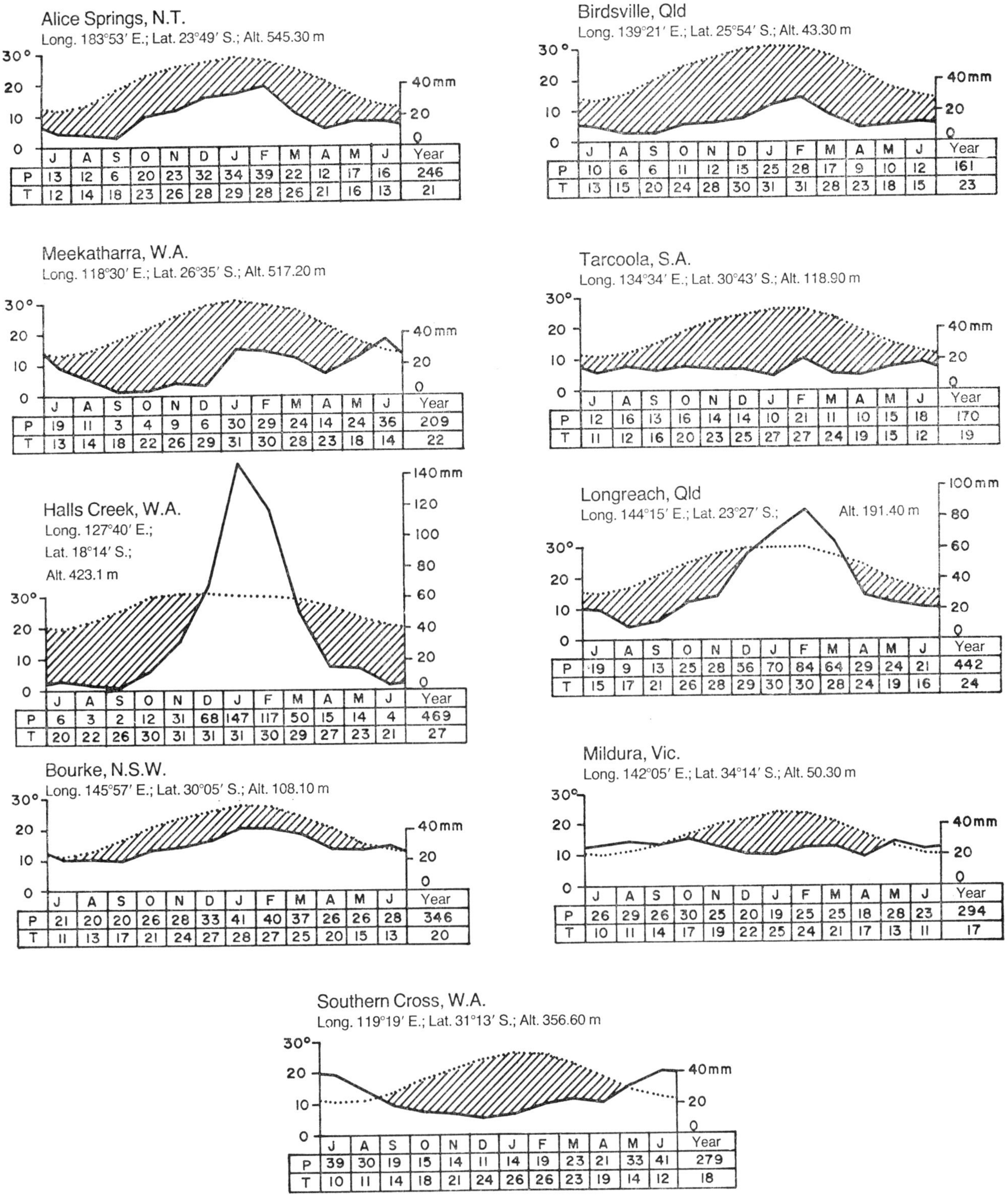

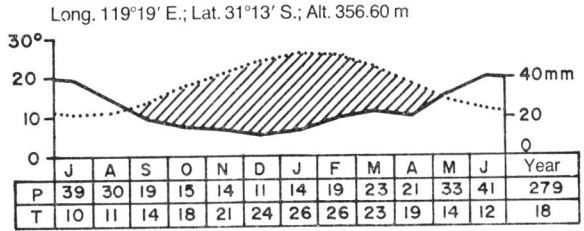

Fig. 6. Schematic representation of the climate at several stations in the Australian arid and semi-arid zone.

Arid hummock grassland is dominated by grasses collectively known as spinifex, consisting of species of the genera *Triodia* and *Plechtrachne*. Grasslands of this type are found in a variety of particularly xeric habitats from sand dunes and sand plains to rocky hillslopes. Typically these grasses occur as individual hummocks of 1-1.5 m diameter and separated by bare areas except following rains when short-lived annuals may form a more complete ground cover. Arid hummock grasslands occur most extensively over the eastern half of Western Australia, northern South Australia and the southern half of the Northern Territory.

Even more widespread than the arid hummock grassland and of roughly comparable total area is the *Acacia* shrubland. The dominant species is most often *Acacia aneura* or mulga. The Acacias of arid Australia are thornless, attain a height usually less than 8 m, and have a characteristic branching close to the ground. Not uncommonly these occur in a distinct banded or grove pattern. Typically occurring with the Acacia are low shrubs, perennial tussock grasses and a wide range of ephemeral herbaceous species following rains. *Acacia* shrubland occurs across the entire arid area from the Indian Ocean to central Queensland, and between the approximate latitudinal limits of 21° and 33° S., i.e. in an area of overlap between summer and winter rainfall (Nix and Austin, 1973).

The shrub steppe formation occurs principally on calcareous and alkaline soils in the cooler portions of the arid area, generally south of latitude 26° S. where there is greater effective rainfall during the winter season. The dominants are low chenopod shrubs of the general *Atriplex* and *Kochia* known commonly as salt-bush and bluebush. Open spaces between the shrubs are commonly bare over long dry periods, but have grasses or annuals following rains.

Arid tussock grasslands occur most widely over the north-eastern portion of the semi-arid area (described below) but are found selectively within that part of the arid area with a well-defined summer rainfall maximum. Typically these grasslands occur on areas with grey cracking clays in small areas of central Australia and over larger areas of south-western Queensland. The dominants are perennial tussock grasses of the genus *Astrebla*. Other perennial grasses are not uncommon, and following periods of rain annual grasses and forbes are also usually present. These grasses are from 0.5 to 1.0 m in height.

Large parts of the Australian arid area support a grazing economy. Productivity and stocking rates differ markedly however. From a commercial grazing point of view, the extensive xerophytic hummock grasslands are the poorest, and the arid tussock grasslands are the most productive. The dominant shrub species of both the *Acacia* shrubland and shrub steppe are palatable to stock, and these form an important feed reserve for livestock over long dry periods. Throughout the arid area there is a continuing problem of serious degradation of grazing land during periods of drought which can extend over several years. Much damage of this kind due to overgrazing occurred in early years of settlement, resulting in serious erosion and loss of productivity on some land types (Condon, 1968). In these times of stress grazing pressures on the palatable perennial grass and shrub species are apt to be excessive in the absence of other available feed, particularly in those areas within access of the limited stock watering points occurring under these conditions. Particularly in these grazing lands, sound management practices in the interests of conservation and natural regeneration are required.

Climatic regime of the semi-arid zone

This area is best considered as transitional from the inner arid core area to the higher rainfall zones closer to the northern north-eastern, south-eastern and south-western coasts. It varies in width from about 200 to 500 km, and as shown in Figure 6 (Halls Creek, Longreach, Bourke, Mildura, Southern Cross), has quite distinctive rainfall and temperature conditions in different sectors. For purposes of description here, the semi-arid area is conveniently divided into three major sectors: northern, eastern, and south-western.

Throughout all of the northern semi-arid lands, from Broome, Western Australia, to northern interior Queensland, rainfall is strongly concentrated in the summer months. Generally over 80 per cent of annual rainfall occurs within four months, December to March (Halls Greek, see Fig. 6). The wettest month on average is January, but rainfall is quite variable in temporal incidence from year to year as well as in total amount. Average variability of annual rainfall ranges from 20 to 35 per cent, being highest along the drier margin and where rainfall is influenced by the erratic occurrence of tropical cyclones. The number of months with mean rainfall less than 30 mm is within the range of four to seven as shown on the map. Rainfall occurring in October and November is mainly from scattered local convectional systems, and with accompanying high temperatures and potential evapotranspiration rates, these early rains are largely ineffective for sustained plant growth. Throughout this northern sector, mean temperatures of the warmest month exceed 30° C, and of the coolest month are in the range of 10° to 20° C. Highest daily maximum temperatures occur just prior to the onset of the summer wet season, these being generally in the range of 35° to 38° C. This sector of the semi-arid area is not in the zone of frost occurrence.

The eastern semi-arid sector not only has a transition to lesser aridity towards the coast, but also a transition in seasonal rainfall regime from strong summer concentration and winter drought in the north-east (Longreach, see Fig. 6) through an area of overlapping summer and winter rainfall in the central east (Bourke, see Fig. 6), to a moderate winter concentration of rainfall in the south-east (Mildura, see Fig. 6). This eastern sector spans a large range of latitude—from the Tropic to about 38° S.—and thus also has a considerable north/south transition in temperature. As seen from the map, mean temperatures of the coldest month are in the range of 10° to 20° C (mild winters) in the north, and between 0° and 10° C (cool winters) in the south. Summer conditions range from very warm (over 30° C) in the north to warm (20° to 30° C) in the south. In the central and southern parts of this sector the interval over which frosts occur on average is from 50 to 150 days in length.

A western extension of the semi-arid lands of eastern Australia is found in South Australia in the Flinders Range and Eyre Peninsula. This is a comparatively small area which has rainfall more strongly concentrated in the winter months, and summers which are distinctly drier than in the semi-arid lands of New South Wales.

The south-west of Western Australia has semi-arid climatic conditions broadly similar to those of South Australia, but with an even more distinctly dry summer period of four to seven months following the winter rains. Erratic summer rains do occur in this area, however, often in association with tropical cyclones that have moved in a general south-easterly direction from the north-west. As is to be expected, winter rainfall diminishes northward and summer rainfall decreases southward across this sector of the semi-arid area. In the northern part mean temperatures in summer are between 20° and 30° C, and winters are mild (10° to 20° C) with few frosts. To the south winters are notably cooler (0° to 10° C) with a frost period on average extending over 150 days in southern inland locations.

Soils, vegetation and land use in semi-arid zones

Many of the soil types identified in the area defined as arid are also found in the semi-arid area. As in the arid area there is commonly a close association of soils and parent materials and between soils and vegetation. In large areas of the south-west of Western Australia, the Eyre Peninsula and southernmost parts of South Australia and in western Victoria, there are solodized solonetz and solodic soils with moderately acid, grey to brown sandy or loamy A horizons above heavy clay subsoils with large prismatic or blocky structural features. Solonized brown soils consisting of calcareous and sandy earths with increasing texture with depth also occur widely. Red earths and red-brown earths are also found in the semi-arid lands of south-western and south-eastern Australia. Widely occurring through the semi-arid areas of northern New South Wales, Queensland and the Northern Territory are grey, brown and red cracking clays with self-mulching surfaces when dry. These soils are characterized by deep cracking upon drying, and commonly have distinctive melon hole or 'gilgai' micro-relief features associated with expansion and contraction of these clays with changing water content.

Vegetation and grazing land characteristics of semi-arid lands in Australia are described in some detail in a comprehensive work edited by Moore (1970). As with soils the vegetation of the semi-arid area is often of the same type as occurs in the true arid climate of the interior. No clearly definable demarcation between arid and semi-arid areas is possible from the occurrence of vegetation formations alone, principally because of the interaction with soil properties, and particularly in respect of their fertility status and water-holding capabilities. For example, large areas in the semi-arid lands of the Northern Territory and northern Queensland have an arid tussock grassland formation as described above.

Extending across northern Australia from the Kimberley area of Western Australia to northern central Queensland there is an ecotone formation known as arid and semi-arid low woodland. This occurs in an intermediate position between arid hummock grasslands or arid tussock grasslands to the south and sub-humid woodlands which occur widely across the higher rainfall areas to the north. This formation consists typically of a mixture of species also found in adjoining areas and with single-stemmed tree components (principally *Eucalyptus* species) of a height usually less than 8 m. The dominant grasses of this formation are species of *Aristida*.

In the semi-arid area of central and southern Queensland and New South Wales a formation known as semi-arid shrub woodland occurs extensively. This has a dominant tree stratum of moderate height and principally of *Eucalyptus* and *Acacia* species, and an understory of low trees or shrubs commonly of the genera, *Eremophila, Acacia,* and *Myoporum.* Grasses of this formation are mainly *Aristida* species in Queensland and northern New South Wales or *Danthonia* and *Stipa* species in southern New South Wales and Victoria. On the more arid side this formation gives way to arid tussock grasslands in the north and to either *Acacia* shrubland or chenopod shrub steppe in the south. On the sub-humid side to the east, the formation gives way to sub-humid woodlands or *A. harpophylla* forest in northern and central Queensland, or to temperate woodlands in New South Wales.

The distinctive shrubland formation called Mallee occupies much of the semi-arid parts of south-western New South Wales, western Victoria, southern South Australia and southern Western Australia. The dominants here are several species of multi-stemmed *Eucalyptus* which usually attain a height not more than 8 m. A sparse ground stratum of low shrubs or hummock grass is also usually present. This formation typically gives way to either *Acacia* shrubland or shrub steppe along the more arid margin.

Land use in the semi-arid area consists of either extensive beef cattle or sheep grazing, the former occurring exclusively in the northern sector, and the latter being dominant in southern areas of New South Wales, western Victoria, South Australia, and Western Australia. In central and southern Queensland and northern New South Wales beef cattle and sheep grazing are both well represented. Stocking rates vary greatly according to the productivity of these grazing lands, ranging generally from 25 to 200 cattle or 200 to 1,600 sheep/1,000 ha.

Land use in the arid and semi-arid areas of Australia has two distinguishing features when compared with that in comparable climates elsewhere. Firstly, Australia, being a comparatively recently settled country, has no long established agricultural traditions, nor identifications sociologically and ethnically as are often found in other countries. Secondly, with the exception of the Murray Valley and its tributaries, irrigated agriculture is of much less significance throughout these arid and semi-arid environments.

Sahara; Sahelian and Sudanese zones

Sahara

This is limited to the north by the Mediterranean steppes and to the south by the Sahelian steppes with summer rains. About ten countries are concerned: Algeria, Egypt, Libya, Mali, Morocco, Mauritania, Niger, Sudan, Chad and Tunisia.

The whole area has a very uniform climate: very low rainfall, less than 100 mm/year, irregular, so drought is almost permanent. The studies of Dubief (1959, 1963, 1971) have shown that in fact there are important climatic distinctions between the eastern Sahara, where the mean annual rainfall is near 0 mm (Aswan, Wadi-Halfa, Helwan, see Fig. 7), and the 'oceanic' Sahara which benefits from markedly higher precipitation and slightly higher air humidity; Atar (Fig. 7) in Mauritania registers an average of 93 mm/year and relative humidity varies between 20 and 40 per cent. There are also temperature differences between north and south. In spite of the high annual means for the whole of the Sahara (generally 20° to 25° C), it is in fact not uncommon for the absolute minimum to fall to between 0° and —5° C in the northern half (In Salah, Fig. 7; Adrar), while it almost never freezes in the south. In the mountains the thermometer often falls below —10° C. Winters are cool (0° to 10° C) in the major Sahara mountain ranges: Hoggar (3,000 m), Tassili Ajjer (2,160 m) and Tibesti (2,400 m); snow may fall there.

In contrast to the generally accepted idea that rainfall is irregular with no seasonal pattern, dominant regional rainfall patterns do exist. The only area with truly aseasonal rainfall is the central and eastern Sahara, especially in Libya and Egypt (Tegerhi, Kufra, Wadi-Halfa).

Natural vegetation

Natural vegetation exists but is very scattered and extremely poor in species. These belong in fact to a small group of families found in other deserts: *Papilionaceae (Acacia, Retama), Zygophyllaceae (Balanites), Polygonaceae (Calligonum), Tamarix, Ziziphus,* etc. A genus of Crucifer (*Schouwia*) has only two species quite characteristic of these parts of Africa, which can constitute large stands with

Helwan (Egypt)
Long. 31°20′ E.; Lat. 29°52′ N.; Alt. 35 m

	J	F	M	A	M	J	J	A	S	O	N	D	Year
P	7	4	5	3	0	0	0	0	0	2	4	5	30
T	13	14	17	21	25	27	28	28	26	24	20	15	21,5

Atar (Mauritania)
Long. 13°04′ W.; Lat. 20°31′ N.; Alt. 225 m

	J	F	M	A	M	J	J	A	S	O	N	D	Year
P	1	1	1	0	2	9	6	31	30	4	6	2	93
T	20	21	25	28	30	34	34	34	33	30	25	21	28

In Salah (Algeria)
Long. 2°29′ E.; Lat. 27°12′ N.

	J	F	M	A	M	J	J	A	S	O	N	D	Year
P	4	3	5	2	1	0	0	1	2	3	4	6	31
T	14	16	17	22	25	34	35	34	32	27	19	14	24

Khartoum (Sudan)
Long. 32°33′ E.; Lat. 15°37′ N.; Alt. 148 m

	J	F	M	A	M	J	J	A	S	O	N	D	Year
P	0	0	0	1	3	10	58	86	18	4	0	0	180
T	23	24	28	32	34	34	32	30	32	32	28	25	29,5

Maiduguri (Nigeria)
Long. 13°10′ E.; Lat. 11°51′ N.; Alt. 354 m

	J	F	M	A	M	J	J	A	S	O	N	D	Year
P	0	0	2	5	42	96	217	238	107	18	0	0	725
T	22	24	29	31	32	30	27	26	27	28	25	22	27

Nouakchott (Mauritania)
Long. 15°56′ W.; Lat. 18°07′ N.; Alt. 5 m

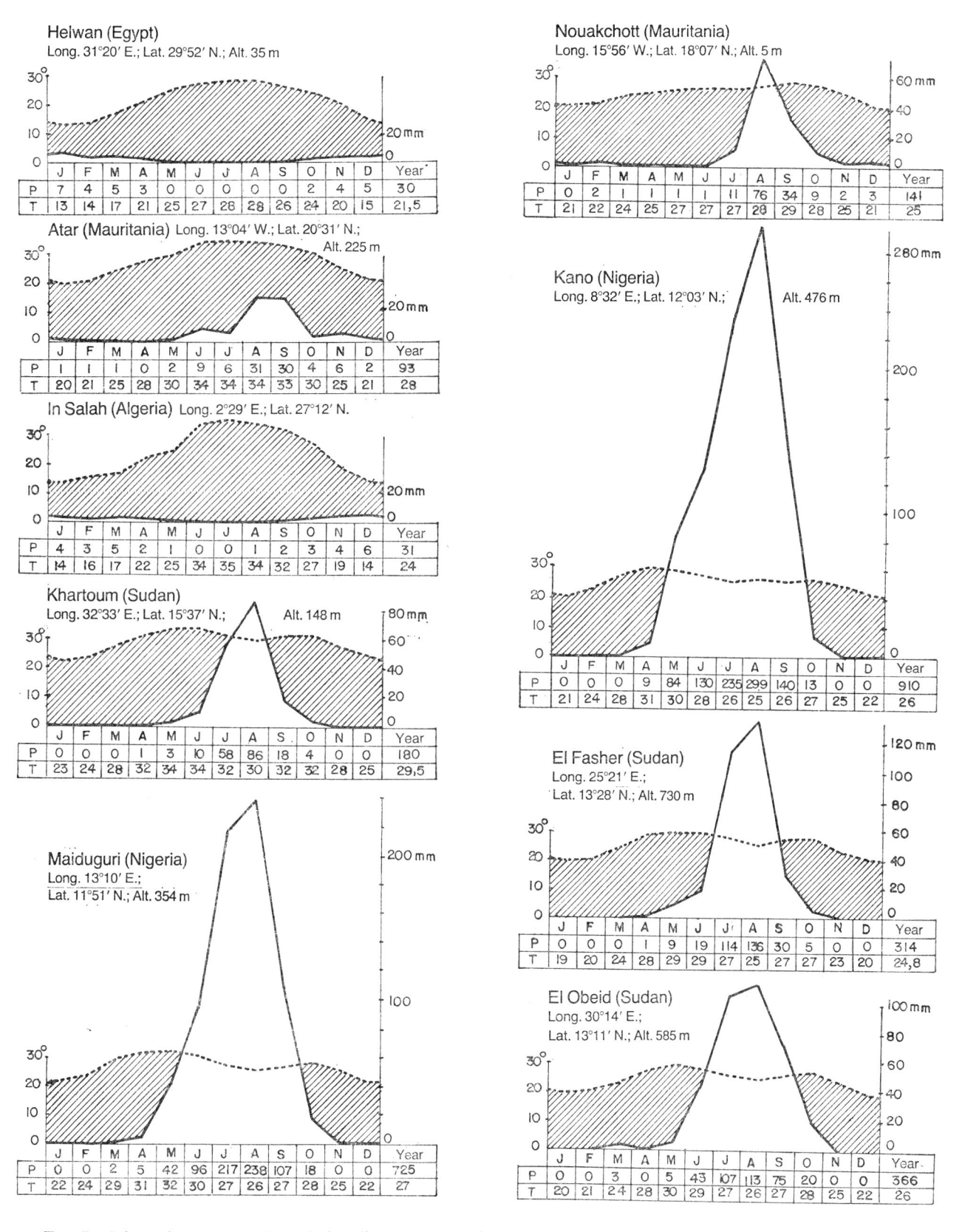

	J	F	M	A	M	J	J	A	S	O	N	D	Year
P	0	2	1	1	1	1	11	76	34	9	2	3	141
T	21	22	24	25	27	27	27	28	29	28	25	21	25

Kano (Nigeria)
Long. 8°32′ E.; Lat. 12°03′ N.; Alt. 476 m

	J	F	M	A	M	J	J	A	S	O	N	D	Year
P	0	0	0	9	84	130	235	299	140	13	0	0	910
T	21	24	28	31	30	28	26	25	26	27	25	22	26

El Fasher (Sudan)
Long. 25°21′ E.; Lat. 13°28′ N.; Alt. 730 m

	J	F	M	A	M	J	J	A	S	O	N	D	Year
P	0	0	0	1	9	19	114	136	30	5	0	0	314
T	19	20	24	28	29	29	27	25	27	27	23	20	24,8

El Obeid (Sudan)
Long. 30°14′ E.; Lat. 13°11′ N.; Alt. 585 m

	J	F	M	A	M	J	J	A	S	O	N	D	Year
P	0	0	3	0	5	43	107	113	75	20	0	0	366
T	20	21	24	28	30	29	27	26	27	28	25	22	26

FIG. 7. Schematic representation of the climate at several stations in Algeria, Egypt, Mauritania, Nigeria and Sudan

occasional moisture. *Panicum turgidum*, a *Gramineae*, is very widespread in these deserts. A good summary of the fauna is found in Monod (1973).

Oases play an essential role for the nomadic and sedentary populations; date palms (*Phoenix dactylifera*) and tiny patches of cereals and vegetables, irrigated by a great variety of mostly ancient techniques, provide essential foodstuffs.

Sahelian and Sudanese zones

These are regions of thorny steppes (Sahel) and of different types of savannah (Sudanese zone) which lie to the south of the hyper-arid Sahara. The countries particularly concerned are Senegal (Trochain, 1940), Mauritania (Adam, 1968a), Niger (Duong Huu Thoi, 1950), Nigeria (Keay, 1959), Chad (Gillet, 1968a, 1968b, 1968c), Sudan (Bégué, 1958, 1971; Obeid and Self El Din, 1970; Radwanski and Wickens, 1967), Somalia, especially the coast north of Mogadiscio, the Danakil area, and lastly the Island of Socotra.

Among the most comprehensive biogeographic works on these regions are those of Aubréville (1949), Monod (1957), Keay (1959) and Schnell (1976).

Climates

Climatic limits drawn from average values are necessarily inaccurate. Monod (1973) makes the following point on this subject:

It should not be forgotten that because of the range of annual variability at a given point, the isohyets which seem so precise on our maps only represent rough approximations: the 100 mm curve in Mauritania which 'on average' passes through Nouakchott and Adrar, in 1951-52 lay 300 km further north, and 300 km further south in 1941-42, giving a difference on the ground of 600 km. Trees do not move according to these episodic fluctuations, but for the nomad and his animals the steppe-desert margin shifts over hundreds of kilometers in latitude.

The 'Sahelian steppes', in the broadest sense, stretch from Nouakchott (Fig. 7) to Port Sudan on the Red Sea passing through Timbuktu and Khartoum (Fig. 7). The semi-arid zone corresponds broadly to the Sudanic wooded savannahs, with stations in north Nigeria (Sokoto; Kano, Fig. 7; Maiduguri, Fig. 7) and in the Sudan (El Fasher and El Obeid, Fig. 7). From a biogeographical point of view, it is quite usual to include the Sahel within a 'Sahel-Somali' area. As the diagrams in Figure 7 indicate, these regions have in common a rainfall regime with a single August or July maximum. In the arid areas, there are generally fewer than thirty rainy days, and the dry season normally lasts eight to eleven months. In the semi-arid areas, there are between thirty and sixty rainy days, and the dry season is six to eight months long.

Natural vegetation

The natural vegetation is a thorny steppe with *Acacia* (*A. raddiana, A. senegal, A. seyal*), *Balanites aegyptiaca, Ziziphus mauritiana*, and such grasses as *Aristida adscencionis, A. funiculata, A. mutabilis*).

In the Sudan, this region is designated 'semi-desert' (*Acacia tortilis, Maerua crassifolia* and *Acacia mellifera, Commiphora* desert scrub). In Somalia the corresponding vegetation is the sub-desert steppe of Keay (1959) (*Aristida* sp., *Schoenefeldia gracilis, Acacia, Commiphora*, etc.).

The Island of Socotra, about 200 km from the Somali coast, should be grouped with the Somali arid region. Its northern part probably receives no more than 150 mm of rain per year and its vegetation, apart from endemic species, is reminiscent of the Sahel (Gwynne, 1968).

The Sahel zone has a nomadic pastoral economy.

In the Sudanese zone, there are permanent villages, savannahs which are regularly burned, a typical cereal agriculture (*Pennisetum, Sorghum, Digitaria, Eleusine*), and agrarian landscapes dominated by fairly typical trees (*Vitellaria paradoxa, Faidherbia albida, Adansonia digitata,* etc.). The dominant natural vegetation is a tree and shrub savannah, which is sometimes even wooded. The tropical tree flora is very diversified: *Isoberlinia doka, I. tomentosa, Monotes kerstingii, Acacia* spp., *Combretum* spp., *Terminalia macroptera, Daniellia oliveri*, etc. The main countries concerned are from west to east: Senegal, Mali, Upper Volta, Ghana, Nigeria, Cameroon, Sudan and lowland Ethiopia. Kenya should also be included in this group. The shores of Lake Turkana are arid. Eastern and southern Kenya and vast areas of Tanzania east of Lake Victoria have climates and vegetation types very much like those of the 'Sudanese zone'.

Southern, south-western Africa and Madagascar

In this part of the world the driest areas are in the west, where the coastal Namib desert is more than 2,000 km long, and also in the extreme south of Madagascar. Aridity decreases from west to east; in western Madagascar it decreases from south to north. It should be noted that the hyper-arid Namib desert benefits from high atmospheric humidity which is not taken into account in the climatic diagrams. It is estimated that here, as in Madagascar, condensation of dew can reach 40 mm per year.

The Kalahari in Botswana, more continental than the Namib, has been mapped in the 'arid' climate group. It is wrong to call this a 'desert', since rainfall is generally between 150 and 300 mm and the natural vegetation consists of trees. The works of Logan (1960, 1969) and the synthesis by Walter (1973) well describe these regions.

Climates

In this part of the world, arid, permanently very hot climates ($t_m > 30°$ C) do not exist; even very hot summers ($t_x > 30°$ C) are practically unknown. The commonest temperature group is certainly that with temperate winters ($10° < t_m < 20°$ C), and hot summers ($20° < t_x < 30°$ C). This climate is represented at the most southerly stations of Port Elizabeth and Cape Town (Fig. 8), on the eastern coast of Mozambique (Maputo, Fig. 8), the continental regions of Botswana (Mahalapye, Fig. 8) and southern Madagascar (Tulear, Fig. 8). Monthly means below 10° C are possible in the eastern uplands (Bloemfontein and Johannesburg, Fig. 8). There is a low average annual temperature range in these hyper-arid regions. At Swakopmund for example (Fig. 8), the July mean is 14° C, and those for February and March vary between 17.5° and 18° C. The normal rainfall pattern is irregular in the hyper-arid regions, winter rains in the south-west (Cape Town), two rainy seasons in the south-east, and summer rains elsewhere.

Natural vegetation

The desert vegetation of the Namib is very sparse, but it is rich in species. These are mostly plants quite characteristic to these regions, including numerous Mesembryanthemums, an extremely diversified family in South Africa. The northern Namib has a strange Gymnosperm, *Welwitschia bainesii* (*W. mirabilis*). Plant geographers often put the Namib and the more southerly Karoo in the same floristic region (Volk 1964, 1966). In both cases the floristic characteristic is probably the great diversification of fleshy-leaved species such as *Zygophyllum* sp., *Aloe* sp., *Asclepiadaceae* and *Crassulaceae* (*Cotyledon, Crassula,* etc.).

The tree steppe vegetation of the Kalahari occurs to the east of the preceding type. The plant landscape is very like that of the Sahel, with a poor savannah scattered with thorny shrubs. *Acacias* are abundant, as are various species of *Apocynaceae* (*Carissa*), *Capparidaceae* (*Cadaba*) and *Combretaceae* (*Combretum* sp.).

The sclerophyllous shrub vegetation, commonly designated 'Cape maquis' corresponds to the dry regions with winter rain. A physionomic analogy (Sclerophyllous brush, 3 to 5 m tall) corresponds to this Mediterranean maquis climatic analogy. The flora of these plant formations (the 'fynbos') is extremely rich and very original (the *Proteaceae* and *Ericaceae* families are important).

A bush with *Didiereaceae* covers most of the arid part of the south of Madagascar. *Didiereaceae* family has two important genera: *Didierea* and *Alluaudia*. The cactus-like Euphorbias can also be codominant (Humbert and Cours Darne 1965; Battistini and Richard Vindard 1972).

North America

United States of America

In the United States of America, there are vast dry regions in the western half of the country. However, apart from some small areas such as the famous 'Colorado desert' and 'Death Valley', aridity does not reach Sahara proportions in the United States. The States most affected by climatic drought are California, Arizona and New Mexico in the south. Large areas of Nevada and Utah receive little

Port Elizabeth (South Africa)
Long. 25°37′ E.; Lat. 33°59′ S.; Alt. 55 m

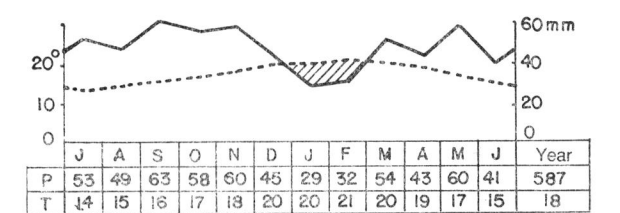

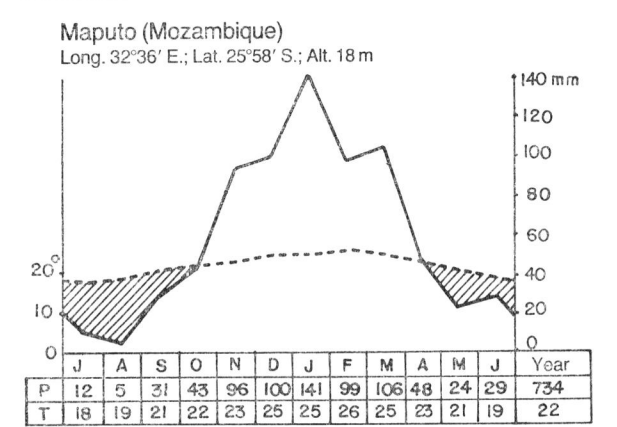

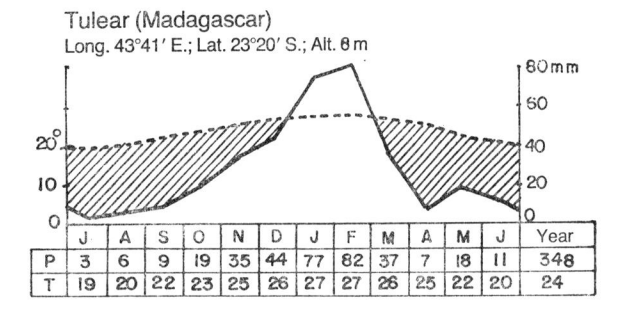

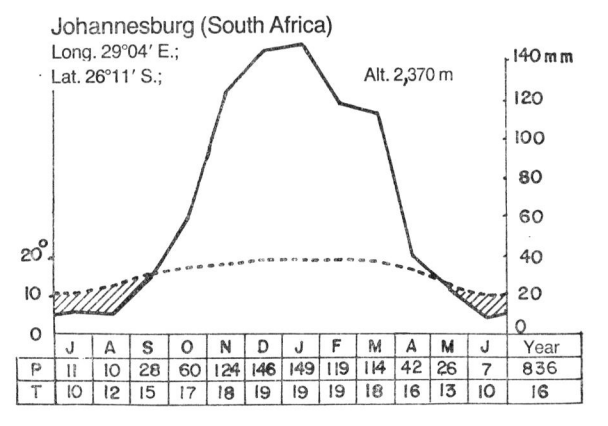

	J	A	S	O	N	D	J	F	M	A	M	J	Year
P	53	49	63	58	60	45	29	32	54	43	60	41	587
T	14	15	16	17	18	20	20	21	20	19	17	15	18

Maputo (Mozambique)
Long. 32°36′ E.; Lat. 25°58′ S.; Alt. 18 m

	J	A	S	O	N	D	J	F	M	A	M	J	Year
P	12	5	31	43	96	100	141	99	106	48	24	29	734
T	18	19	21	22	23	25	25	26	25	23	21	19	22

Tulear (Madagascar)
Long. 43°41′ E.; Lat. 23°20′ S.; Alt. 8 m

	J	A	S	O	N	D	J	F	M	A	M	J	Year
P	3	6	9	19	35	44	77	82	37	7	18	11	348
T	19	20	22	23	25	26	27	27	26	25	22	20	24

Johannesburg (South Africa)
Long. 29°04′ E.; Lat. 26°11′ S.; Alt. 2,370 m

	J	A	S	O	N	D	J	F	M	A	M	J	Year
P	11	10	28	60	124	146	149	119	114	42	26	7	836
T	10	12	15	17	18	19	19	18	16	13	10	16	

Cape Town (South Africa)
Long. 18°29′ E.; Lat. 33°56′ S.; Alt. 12 m

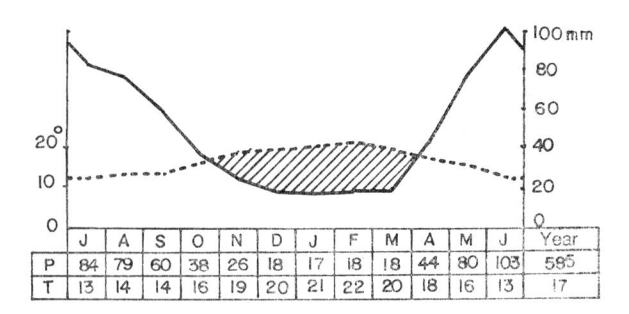

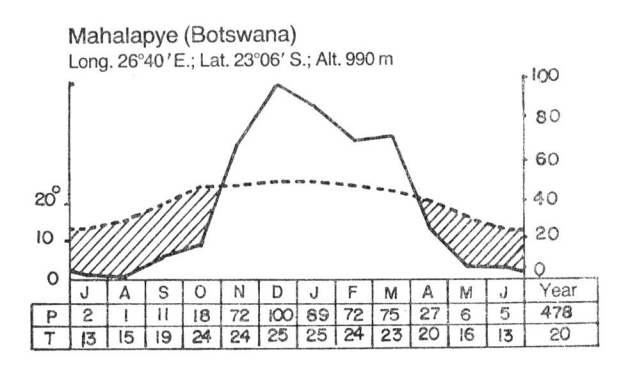

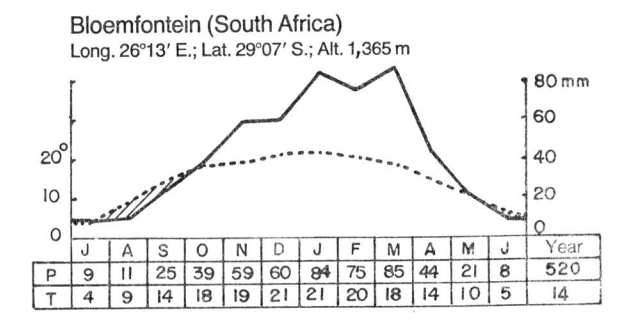

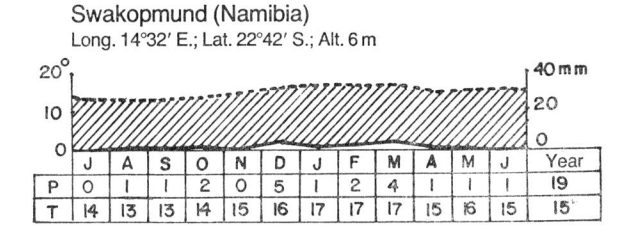

	J	A	S	O	N	D	J	F	M	A	M	J	Year
P	84	79	60	38	26	18	17	18	18	44	80	103	585
T	13	14	14	16	19	20	21	22	20	18	16	13	17

Mahalapye (Botswana)
Long. 26°40′ E.; Lat. 23°06′ S.; Alt. 990 m

	J	A	S	O	N	D	J	F	M	A	M	J	Year
P	2	1	11	18	72	100	89	72	75	27	6	5	478
T	13	15	19	24	24	25	24	24	23	20	16	13	20

Bloemfontein (South Africa)
Long. 26°13′ E.; Lat. 29°07′ S.; Alt. 1,365 m

	J	A	S	O	N	D	J	F	M	A	M	J	Year
P	9	11	25	39	59	60	84	75	85	44	21	8	520
T	4	9	14	18	19	21	21	20	18	14	10	5	14

Swakopmund (Namibia)
Long. 14°32′ E.; Lat. 22°42′ S.; Alt. 6 m

	J	A	S	O	N	D	J	F	M	A	M	J	Year
P	0	1	1	2	0	5	1	2	4	1	1	1	19
T	14	13	13	14	15	16	17	17	17	15	16	15	15

FIG. 8. Schematic representation of the climate at several stations of South Africa, Botswana, Madagascar, Mozambique, and Namibia.

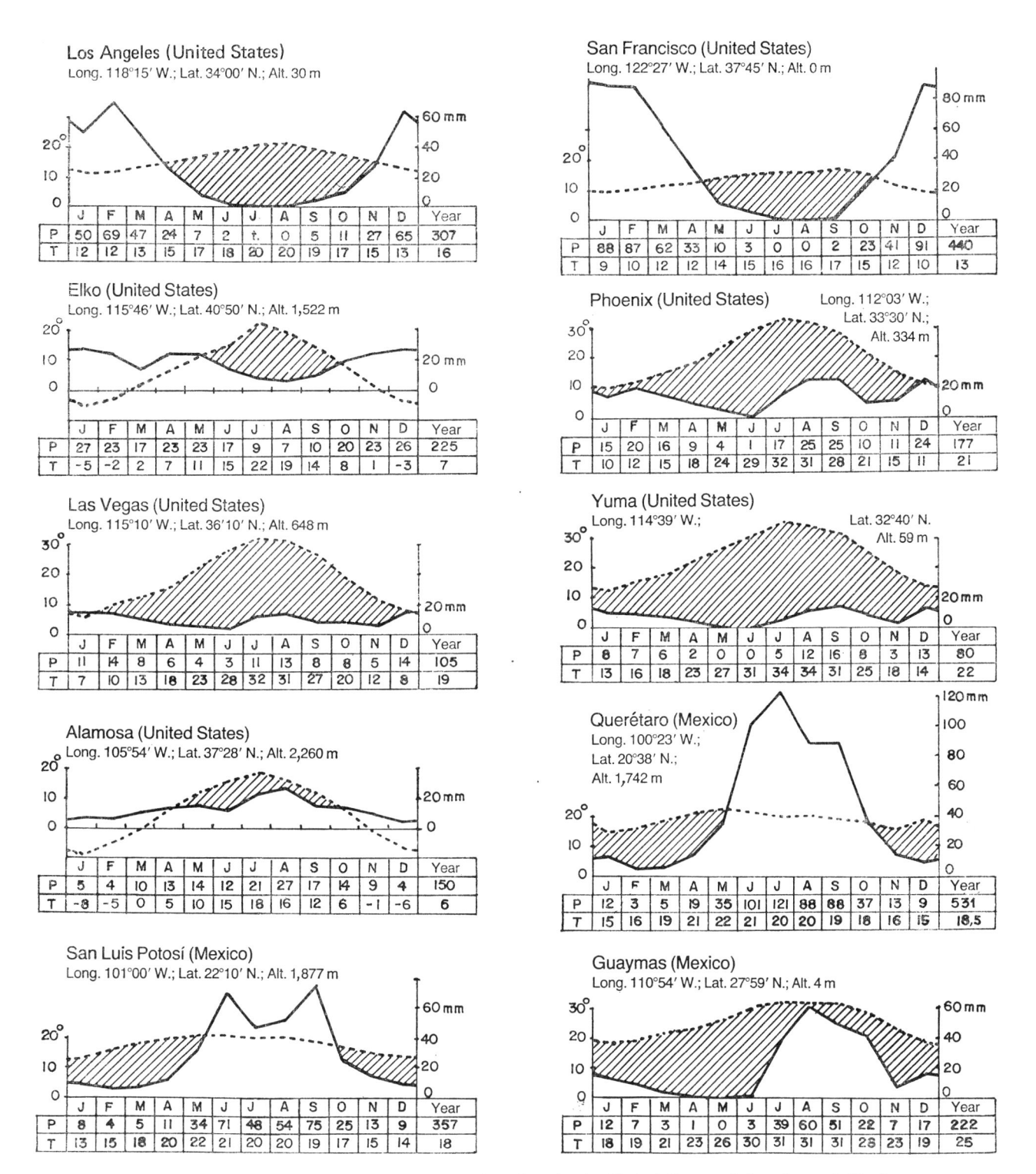

Los Angeles (United States)
Long. 118°15' W.; Lat. 34°00' N.; Alt. 30 m

	J	F	M	A	M	J	J	A	S	O	N	D	Year
P	50	69	47	24	7	2	t.	0	5	11	27	65	307
T	12	12	13	15	17	18	20	20	19	17	15	13	16

Elko (United States)
Long. 115°46' W.; Lat. 40°50' N.; Alt. 1,522 m

	J	F	M	A	M	J	J	A	S	O	N	D	Year
P	27	23	17	23	23	17	9	7	10	20	23	26	225
T	-5	-2	2	7	11	15	22	19	14	8	1	-3	7

Las Vegas (United States)
Long. 115°10' W.; Lat. 36°10' N.; Alt. 648 m

	J	F	M	A	M	J	J	A	S	O	N	D	Year
P	11	14	8	6	4	3	11	13	8	8	5	14	105
T	7	10	13	18	23	28	32	31	27	20	12	8	19

Alamosa (United States)
Long. 105°54' W.; Lat. 37°28' N.; Alt. 2,260 m

	J	F	M	A	M	J	J	A	S	O	N	D	Year
P	5	4	10	13	14	12	21	27	17	14	9	4	150
T	-8	-5	0	5	10	15	18	16	12	6	-1	-6	6

San Luis Potosí (Mexico)
Long. 101°00' W.; Lat. 22°10' N.; Alt. 1,877 m

	J	F	M	A	M	J	J	A	S	O	N	D	Year
P	8	4	5	11	34	71	48	54	75	25	13	9	357
T	13	15	18	20	22	21	20	20	19	17	15	14	18

San Francisco (United States)
Long. 122°27' W.; Lat. 37°45' N.; Alt. 0 m

	J	F	M	A	M	J	J	A	S	O	N	D	Year
P	88	87	62	33	10	3	0	0	2	23	41	91	440
T	9	10	12	12	14	15	16	16	17	15	12	10	13

Phoenix (United States)
Long. 112°03' W.; Lat. 33°30' N.; Alt. 334 m

	J	F	M	A	M	J	J	A	S	O	N	D	Year
P	15	20	16	9	4	1	17	25	25	10	11	24	177
T	10	12	15	18	24	29	32	31	28	21	15	11	21

Yuma (United States)
Long. 114°39' W.; Lat. 32°40' N.; Alt. 59 m

	J	F	M	A	M	J	J	A	S	O	N	D	Year
P	8	7	6	2	0	0	5	12	16	8	3	13	80
T	13	16	18	23	27	31	34	34	31	25	18	14	22

Querétaro (Mexico)
Long. 100°23' W.; Lat. 20°38' N.; Alt. 1,742 m

	J	F	M	A	M	J	J	A	S	O	N	D	Year
P	12	3	5	19	35	101	121	88	88	37	13	9	531
T	15	16	19	21	22	21	20	20	19	18	16	15	18,5

Guaymas (Mexico)
Long. 110°54' W.; Lat. 27°59' N.; Alt. 4 m

	J	F	M	A	M	J	J	A	S	O	N	D	Year
P	12	7	3	1	0	3	39	60	51	22	7	17	222
T	18	19	21	23	26	30	31	31	31	28	23	19	25

FIG. 9. Schematic representation of the climate at several stations in the United States of America and Mexico.

rainfall, as do southern Idaho and eastern Oregon and Washington state. Stoeckeler (1970) contains a good bibliography, with nearly 200 references on the dry regions of this part of the world.

Climates

Figure 9 gives a schematic representation of the climate at several stations in the United States of America and in Mexico. The considerable span in latitude (from about 33° to 45° N.), in longitude (between the 100° W. and the Pacific), and in altitude (parts of the Rocky Mountains and the Sierra Nevada), explains the great diversity of climate. A detailed analysis of the principal climatic regions has been made by Thornthwaite (1941, 1948). All the types of dry climate possible at these non-tropical latitudes exist in the United States. The driest or hyper-arid regions, or 'deserts', are the Great Basin, Mojave, Colorado (Sonoran Desert: station at Yuma, Fig. 9) and Chihuaha, which is mostly in Mexico.

There are several criteria for regional distinctions (see Fig. 9): (a) regions with temperate winters ($10° < t_m < 20°$ C); summers can be very hot (32° C at Phoenix in Arizona) or hot (20° to 22° C in July-August at San Diego and Los Angeles); (b) regions with cool winters ($0° < t_m < 10°$ C); summers can be very hot (more than 32° C in June at Las Vegas), hot or temperate (16° C in September at San Francisco); (c) regions with cold winters ($t_m < 0°$ C); according to altitude, summers are temperate (18° C in July at Alamosa, altitude 2,260 m) or hot (stations at Grand Junction, Pocatello, Elko).

Rainfall regimes are very diverse and all possible cases are represented, from spring and summer rain east of the Rockies, to Californian types with very marked summer drought.

Rainfall also varies enormously from the arid lowlands of Nevada (105 mm/year at Las Vegas) and of Arizona (80 mm/year at Yuma), to the very wet regions in the west of Washington and Oregon states (1,930 mm/year at Astoria on the Pacific Coast).

Natural vegetation and land use

The U.S. Department of Agriculture (1941, 1949, 1955, 1957, 1958) has published important maps of forests and soils and their uses. An excellent general paper on xerophytes has been done by Shreve (1942). Shreve and Wiggins in 1964 published *Vegetation and Flora of the Sonoran Desert*. The flora by Standley (1920-26) is still very useful for woody species; *Desert Wild Flowers* by Jaeger (1956) should also be mentioned.

In the Great Basin Desert east of San Francisco, steppe vegetation predominates, but is poor in species: mainly *Artemisia tridentata* and *Atriplex confertifolia*. Further south, in the Mojave Desert which is almost at the same latitude as Los Angeles, the flora is enriched by *Yucca schidigera, Larrea divaricata, Franseria dumosa*. Even further south, in the Sonoran desert which includes notably the peninsula of Baja California (Mexico) and the Lower Colorado Valley (Yuma region), sub-tropical temperatures are accompanied by a different flora, characterized mainly by the abundance of columnar or candelabra-shaped cacti: *Carnegia gigantea,* which grows over 15 m tall, is probably the commonest srecies. Mesquite (*Prosopis juliflora*) and *Acacia willardiana,* the only American *Acacia* with phyllodes, are also widespread. Finally, east of the Sonoran desert, the essentially Mexican Chihuahua desert is much more diverse than the former. Its average altitude varies between 1,000 and 2,000 m, and temperatures are markedly lower. A thorny steppe is still present but numerous perennial grasses give this landscape an aspect intermediate between a steppe and a wooded savannah. Monod (1973) gives interesting comments on the fauna.

Most dry regions of the west are devoted to extensive livestock raising, with complementary crops such as maize. The hyper-arid and arid enclaves of Nevada and Arizona are not suitable however for livestock or agriculture. In contrast, very lucrative irrigated crops are grown in California, Oregon and Washington. Land use is fully covered in the U.S. Department of Agriculture *Yearbooks* (1949, 1955, 1957, 1958).

A great deal of research has been done on the techniques and the species best adapted for reafforestation in various conditions of drought and altitude. These works are listed in Stoeckeler (1970). *Pinus ponderosa* is unquestionably the most widely used species. *P. coulteri* is planted in the lowlands but is a less valuable species. *Pinus contorta* gives good results in the Rocky Mountains.

Mexico

On the whole, this country has predominantly dry climates, except for some mountainous areas, a part of the western coast and the Gulf of Mexico, and Yucatan and Chiapas. According to Contreras Arias (1955), semi-arid land covers 33.4 per cent, and arid land 18.8 per cent, making a total of 52.2 per cent of the whole country.

The driest regions are in the north, especially around the Californian Gulf, in the states of Sonora and Chihuahua. The works of Garcia (1964) and Puig (1976) are among the most comprehensive ecological and climatic works on Mexico. In addition Mosino (1974) and Benassini (1974) give good summaries of climate and hydrology.

Climates

The arid regions of Mexico (Fig. 9) are a southern extension of the dry climates of the United States. The Sonoran desert overlaps into Mexico, on either side of the Gulf of California, and the Chihuaha desert extends far into the centre of the Mexican plateau. It is therefore possible to group these with southern New Mexico and south-western Texas. Small isolated areas of reputedly very arid climate are noted further south by Rzedowski (1973) in the Mezquital valley (state of Hidalgo), and in the regions of Tehuacan (Puebla) and Cuicatlan (Oaxaca).

Two distinct differences from the dry climates of the western United States should be noted: (a) in Mexico there are hardly any arid or semi-arid regions with cold winters ($t_m < 0°$ C); (b) there are coastal climates, mainly semi-arid, with hot winters (t_m: $20°$ to $30°$ C), in Yucatan (station at Merida) and on the coast of the Gulf of California (Mazatlan region).

The most frequent temperature regimes are those with temperate winters ($10° < t_m < 20°$ C) and hot summers ($20° < t_x < 30°$ C); these are mainly found in the centre at approximately 2,000 m altitude (Queretaro, San Luis Potosi, Fig. 9). Further north, or at altitudes of 2,500 m or more, mean winter temperatures are often below $10°$ C. However some regions like Guaymas (Fig. 9) have very hot summers despite their temperate winters.

Rainfall patterns are less varied than in the United States. In Mexico winter rainfall regimes with maximum summer drought, and irregular regimes with erratic rainfall, are practically non-existent. The dominant rainfall pattern around the Gulf of California has a double dry season with winter and summer rains. Further south a tropical regime with a summer maximum predominates.

Natural vegetation

In the driest parts there is a steppe vegetation which is either grassy (Zacatal) with *Andropogon barbinoides, Aristida adscencionis, Bouteloua* sp., and *Liliaceae* spp., or else with thorny or succulent shrubs. A *Zygophyllaceae* (*Larrea divaricata*), which is very common in North America, often dominates the sub-desert steppes. Thicker and floristically very varied stages also exist in the Mexican arid habitats, known by the name of 'matorrales'. These are open scrub with thorny xerophytes and succulents: *Acacia, Opuntia, Yucca, Agave, Myrtillocactus geometrizans, Lemaireocactus dumortieri, Prosopis juliflora, Fouquieria splendens*.

The differences between the species of the Sonoran and Chihuahua deserts are probably due to the distinctly higher temperatures in Sonora. However the history of the vegetation has certainly played an important part in defining the regional components.

The arid and very arid regions of Mexico have not been agriculturally developed except in the narrow coastal area of the Gulf of California. With a few exceptions, Mexico is not a country of irrigated crops and most of the dry areas are devoted to livestock raising.

South America

As in southern Africa, the western part of South America is the most affected by drought. The main cause is in fact the same in both cases: the existence of cold sea currents, the Benguela for the Namib, and the Humbolt for the Atacama. However in South America the western region becomes humid south of Valparaiso, whereas the eastern region, especially Patagonia, is very dry.

Relatively small areas have also been mapped as semi-arid and sub-humid in Colombia, Venezuela (see Ewell and Madriz, 1968) and Brazil (Aubréville, 1961). The work of de Martonne (1935) has become a classic of general climatology. In biogeography, a good overall view is given in the map and notes by Hueck (1972). Other notable works are those of Morello (1955, 1956), Tricart (1966, 1969*a*, 1969*b*) and Cabrera (1971) for Argentina; of Reparaz (1958), Tosi (1960) and Malleux (1975) for Peru; of Schmithüsen (1956) and of di Castri (1968) for Chile; and, for the geomorphology of the Andes, the analysis by Dollfus (1973).

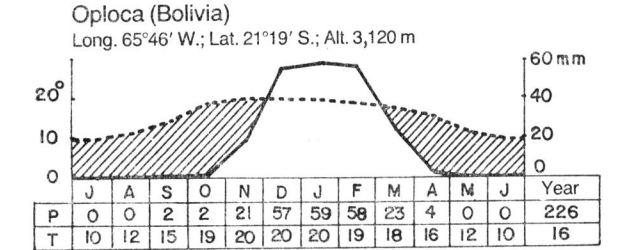

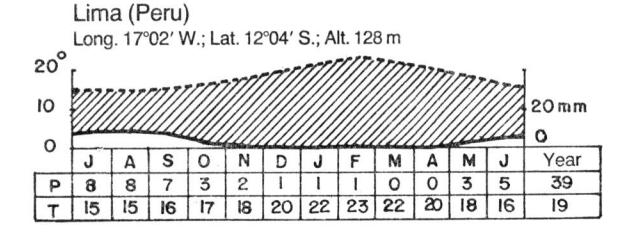

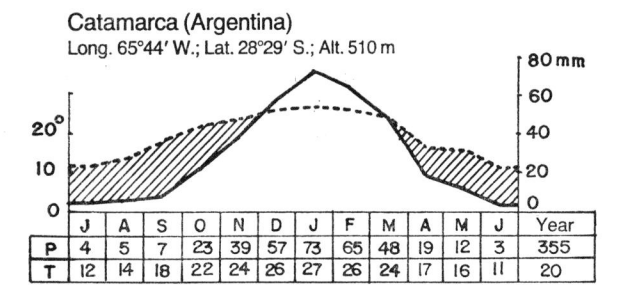

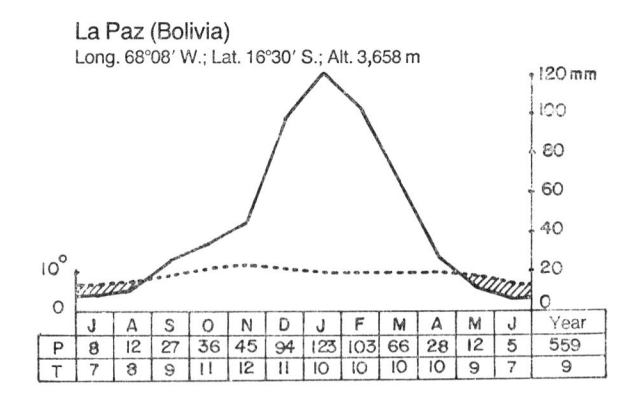

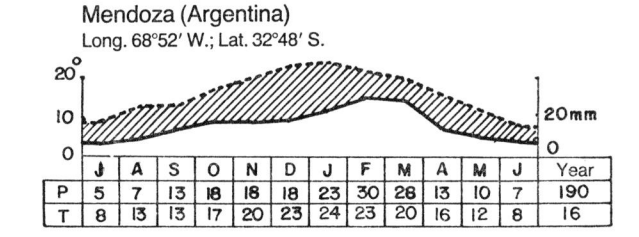

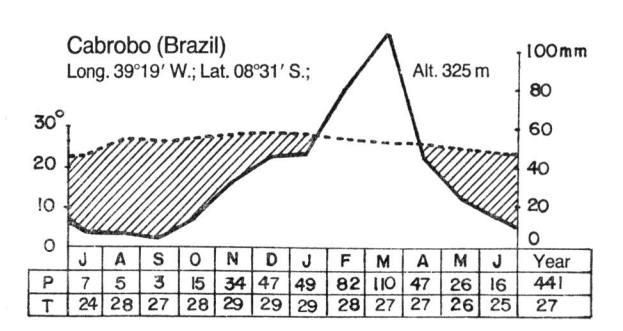

FIG. 10. Schematic representation of the climate at several stations in Argentina, Bolivia, Brazil, Chile and Peru.

Climates

One of the unique features of this huge continent, which stretches diagonally over 5,000 km from Colombia to Patagonia, is that pronounced drought is found as far south as latitude 50°. Puerto Santa Cruz (Fig. 10), in southern Argentina at 50° S., has an arid climate (193 mm/year and ten to twelve dry months); winters are cool, almost cold (January average 1.5° C), summers temperate ($t_x = 14.4°$ C in July). On the other hand, arid regions are found at high altitudes in the Andes, usually between 3,000 and 4,500 m. These are the *punas* which cover particularly large areas in Bolivia. In these *altiplanos*, the distribution of climates is much complicated by numerous sheltered basins dominated by volcanic uplands. It is possible to distinguish according to the degree of aridity (see Fig. 10): (a) the humid puna regions (La Paz), where livestock raising, cereals and potatoes are possible; (b) the semi-arid puna regions (100 to 400 mm/year), used by nomadic llama herders, for example around Lake Poopo, Salar de Uyuni and Oploca; (c) some very sheltered enclaves even have desert punas where the mean annual rainfall does not exceed 100 mm/year (Potrerillos). These high-altitude dry regions have cool winters ($0° < t_m < 10°$ C or $t_m \geqslant 10°$ C) and temperate summers ($10 < t_x < 20°$ C).

The hyper-arid regions lie on the coast of Peru and Chile from the Sechura desert to south of Salar d'Atacama. This includes the Atacama desert proper, containing from north to south the stations of Cartavio, Lima (Fig. 10), Mollendo, Arica, Equique, Antofagasta. These last three stations have a mean annual rainfall below 10 mm. Nevertheless, as in the African Namib, frequent coastal fogs create high atmospheric humidity, especially in winter (from May to September).

Lastly, another arid area has been mapped in Argentina, south of the Tropic of Capricorn, on the western foothills of the Andes across to Patagonia. This contains the stations of San Juan, Mendoza (Fig. 10), Chipoletti (Fig. 10), Santa Maria, etc. There is generally between 100 and 200 mm of rain per year. Winters are cool ($0 < t_m < 10°$ C), summers hot or temperate in the south ($10 < t_x < 20°$ C).

Natural vegetation and land use

In the hyper-arid regions of Peru and northern Chile, the coastal area which benefits from oceanic influences is more favourable than the more continental areas where patches of absolute desert are not rare. It is therefore near the coast that several natural vegetation types are found, in the form of very open steppes with various species of *Cactaceae*, *Bromeliaceae* and Ephemerophytes. Some bushy formations with *Prosopis* are found along watercourses. Where irrigation is possible, there is intensive cropping (cotton, market gardening); this explains why the country is divided into small inhabited areas separated by vast empty areas. According to the land use map of Peru (Zamora, 1971), the total area unfit for agriculture or forestry is 38 per cent of the country; a high proportion of this area is in fact in the hyper-arid zone.

The high-altitude arid regions contain the *punas*, low, open shrubby steppes with *Graminaceae* (*Festuca*), columar cacti, numerous *Compositeae* and *Solanaceae*. The candelabra cacti (*Oreocereus*) are found at 3,500 m altitude. Some meadows of *Pennisetum chilense* and *Festuca scirpifolia* degrade to grassy steppes.

The arid and semi-arid regions of the pediments in Argentina also contain shrubby steppes with a summer rainfall regime in the north (Catamarca, Fig. 10), and a winter one in the south (Chipoletti, Fig. 10). These are the 'shrubby foothill steppes' with *Larrea* (*L. divaricata*, *L. cuneiformis*, *L. nitida*), *Prosopis* (*P. alpataco*, *P. strombulifera*), etc. These plant formations are also found in part of the Argentinian Chaco.

As in Chile and Peru, saline depressions contain halophytes of the genera *Atriplex*, *Suaeda* and *Salicornia*. In these 'monte' regions, cereal crops, and sheep and cattle raising are widespread.

Patagonia is mostly covered by shrubby grass steppes with *Stipa* (*S. patagonica*, *S. humilis*, *S. chrysophylla*), *Festuca*, *Poa*, *Bromus*, etc. The wet meadows contain rich stands of *Cyperaceae* (*Carex gayana*, *C. nebularum*), *Joncaceae* (*J. lesueurii*), etc. The wind and the dryness make agriculture impossible, and these are regions of extensive sheep raising.

The northern tropical countries (Colombia, Venezuela and Brazil) do not have many arid areas. In Colombia, the Santa Marta region (Guajira Peninsula) receives less than 200 mm/year in places; the vegetation includes the 'matorral desertico' and the 'monte espinoso tropical' on the map of the Agustin Codazzi Geographical Institute. These are regions of more or less dense thorny scrub with leguminous plants (*Prosopis*, *Caesalpinia*, *Cassia*), *Capparidaceae*, *Rubiacea* and *Cactaceae*. Goats and sheep are widespread; cereal crops and cotton are grown. Comparable plant formations are found in the east in similar Venezuelan climates, and on the islands near the coast. In Brazil, the driest climates of the country correspond to the *caatingas,* low, thorny forests of very variable density, with Cacti (*Cereus gounellei, C. jamacaru, C. squamosus,* etc.). There are many endemic species. These are pastoral regions, which are also suitable for various crops (cotton, sisal) and, in the valleys, carnauba palm trees (*Copernicia cerifera*), whose leaves are collected for their wax. Average climates are not very dry. Cabrobo (Fig. 10), on the Sao Francisco, is one of the driest areas. The main problem in north-east Brazil is the extreme irregularity of rainfall.

Bibliography[1]

ADAM, J. G. 1968a. La Mauritanie. In: *Conservation of vegetation in Africa south of the Sahara*. I. Hedberg and O. Hedberg (Eds), p. 49-51. Acta Phytogeographia Suecica 54. University of Uppsala, Uppsala.

ADAM, J. G. 1968b. Le Sénégal. In: *Conservation of vegetation in Africa south of the Sahara*. I. Hedberg and O. Hedberg (Eds), p. 65-69. Acta Phytogeographica Suecica 54. University of Uppsala, Uppsala.

ALLUE ANDRADE, J. L. 1966. *Subregiones fitoclimáticas de España*. Instituto Forestal de Investigaciones y Experiencias, Madrid.

ALLUE ANDRADE, J. L.; NAVARRO GARNICA, M. 1970. Spain. In: *Afforestation in arid zones*. R. N. Kaul (Ed.), p. 21-36. Monographiae Biologicae No. 20. W. Junk, The Hague.

AMIRAN, D. H. K. 1964. Land use in Israel. In: *Land use in semi-arid Mediterranean climates/Utilisation des terres en climat semi-aride méditerranéen*, p. 101-112. Arid Zone Research/Recherches sur la zone aride 26. Unesco, Paris.

ANON. 1973. Bibliography on "plant ecology" in Afghanistan. *Excerpta Botanica*, Section B, Band 12, p. 310-315. Gustav Fischer Verlag, Stuttgart.

AUBERT, G. 1951. Les sols des régions semi-arides d'Afrique et leur mise en valeur. In: *Les bases écologiques de la régénération de la végétation des zones arides*, p. 11-25. International Union of Biological Sciences (IUBS), Paris.

AUBREVILLE, A. 1949. *Climats, forêts et désertification de l'Afrique tropicale*. Société d'Editions Géographiques, Maritimes et Coloniales, Paris.

AUBREVILLE A. 1961. Etude Ecologique des principales formations végétales du Brésil, et contribution à la connaissance des forêts de l'Amazonie brésilienne. Centre Technique Forestier Tropical (CTFT), Nogent-sur-Marne.

AUSTRALIA. DEPARTMENT OF NATIONAL DEVELOPMENT. 1970. *Atlas of Australian resources*. (Second Series). Canberra.

AUSTRALIA. BUREAU OF METEOROLOGY. 1974. *Climate of Australia*. Extract from Official Year Book of Australia No. 60, 1974. Government Printing Office, Canberra.

BAGNOULS, F.; GAUSSEN, H. 1957. Les climats biologiques et leur classification. *Annls Geogr.*, 64 (355), p. 193-220.

BARRY, J. P.; CELLES, J. C.; FAUREL, L. 1963 et 1973. *Carte internationale du Tapis végétal et des conditions écologiques; feuilles «Gardaia» à 1/250 000 et «Alger» à 1/1 000 000*. ICITV, Faculté des Sciences, Toulouse.

BATTISTINI, R.; RICHARD-VINDARD, G. (Eds). 1972. *Biogeography and ecology in Madagascar*. Monographiae Biologicae No. 21. W. Junk, The Hague.

BHARADWAJ, O. P. 1961. The arid zone of India and Pakistan. In: *History of land use in arid region*, p. 143-174. Arid Zone research 17. Unesco, Paris.

BEGUE, L. 1958. Les forêts de la République du Soudan. *Bois For. Trop.*, 62, p. 3-20.

BEGUE, L. 1971. Retour au Soudan, *Bois For. Trop.*, 98, p. 3-11.

BENASSINI, O. 1974. Los recursos hidraulogicos de Mexico. In: *El escenario geographico*, p. 173-302. Inst. Nac. Antrop. Hist., Mexico, D.F.

BLASCO, F. 1975. Mangroves of India. *Inst. fr. Pondichéry, Trav. Sect. scient. techn.*, 14, p. 1-175.

BLASCO, F.; LEGRIS, P. 1973. Originalité des climats secs du sud de l'Inde. *Annls Géogr.*, 450, p. 129-150.

BOYKO, H. (Ed.). 1966. *Salinity and aridity. New approaches to old problems*. Monographiae Biologicae No. 16. W. Junk, The Hague.

BURBIDGE, N. T. 1960. The phytogeography of the Australian region. *Aust. J. Bot.*, 8, p. 75-212.

CABRERA, A. L. 1971. Fitogeografia de la Republica Argentina. *Bol. Soc. argent Bot.*, 14, p. 1-42.

CASTRI, F. di. 1968. Esquisse écologique du Chili. In: *Biologie de l'Amérique australe*. C. Delamare Deboutteville and E. Rapoport (Eds), 4, p. 7-52. Centre National de la Recherche Scientifique (CNRS), Paris.

CHAMPION, H. G.; SETH, S. K. 1968. *A revised survey of the forest types of India*. Government of India, Manager of Publications, New Delhi.

CHOUCHANI, B. 1972. *Le Liban, contribution à son étude climatique et phytogéographique*. Doctorat de spécialité. Université Paul Sabatier, Toulouse.

CHRISTIAN, C. S. 1964. Methods of land use in Australia's arid and semi-arid areas. n: *Land use in semi-arid mediterranean climates/Utilisation des terres en climat semi-aride mediterranéen*, p. 117-122. Arid Zone Research/Recherches sur la zone aride 26. Unesco, Paris.

CONDON, R. W. 1968. Estimation of grazing capacity on arid grazing lands. In: *Land evaluation*. G. A. Stewart (Ed.), p. 112-124. Macmillan, Melbourne.

---

1. This bibliography concerns the Explanatory note only. It is limited to the main works on natural vegetation and land use. Some works such as those of Planhol, Ionesco, Hills, Le Houerou, etc., are cited purposely, because of the considerable number of bibliographical references on arid areas they contain. The Unesco series 'Arid zone research' also includes a wealth of information on arid zones.

CONTRERAS ARIAS, A. 1955. In: *Problemas de las zonas aridas de México*. Instituto Mexicano de Recursos Naturales Renovables. Biblioteca Central de la Ciudad Universitaria, Mexico, D.F.

CUNYS H. 1961. *Les déserts dans le monde*. Payot, Paris.

DEMANGEOT, J. 1972. *Les milieux naturels désertiques*. Cours de Géographie physique. Centre de Documentation Universitaire — Société d'Edition d'Enseignement Supérieur (CDU-SEDES), Paris.

DEPOIS, J. 1964. L'utilisation du sol dans les montagnes de l'Atlas. In: *Land use in semi-arid Mediterranean climates/ Utilisation des terres en climat semi-aride méditerranéen*, p. 69-74. Arid Zone Research/Recherches sur la zone aride 26. Unesco, Paris.

DHIR, R. D. 1952. Hydrological research in the arid and semi-arid regions of India and Pakistan. In: *Reviews of research on arid zone hydrology*, p. 96-127. Arid Zone Research 1. Unesco, Paris.

DOBREMEZ, J. F.; VIGNY, F.; WILLIAMS, L. H. J. 1972. *Bibliographie du Népal*. Vol. 3 (Sciences Naturelles), Tome 2 (Botanique). CNRS, Paris.

DOLLFUS, P. 1973. La Cordillère des Andes, présentation des problèmes morphologiques. *Revue Géogr. phys. Géol. dyn.*, 15, p. 157-176.

DRESCH, J. 1968. Reconnaissance dans le Lut méridional. *Bull. Assoc. Géogr. français*, 362-363, p. 143-153.

DRESCH, J. 1972. Sur les relations entre les hautes montagnes et leur piémont en régions arides: les exemples du Liban et du Lut. *Bull. Assoc. Géogr. français*, 399, p. 251-260.

DUCHAUFOUR, P. 1960. *Précis de pédologie*. Masson, Paris.

DUBIEF, J. 1959. *Le climat du Sahara. I. Les températures*. Inst. Rech. sahar., Alger. (Mémoire h.s.).

DUBIEF, J. 1963. *Le climat du Sahara. II. Les précipitations*. Inst. Rech. sahar., Alger. (Mémoire h.s.).

DUBIEF, J. 1971. Die Sahara, eine Klima-Wüste. In: *Die Sahara und ihre Randgebiete*, I. H. Schiffers (Ed.), p. 227-348. München.

DUONG HUU THOI. 1950. Etude préliminaire de la végétation du delta central nigérien. In: *Proc. 2e Conf. intern. Africanistes occid.* (Bissau, 1947), 2 (1), p. 53-156. Junta Invest. colon. Lisboa.

DUPIAS, G. et al. 1965. *Carte de la végétation de la France à 1/200,000, Corse, avec Notice Explicative*. Centre National de Recherche Scientifique (CNRS), Paris.

DURRIEU, G. 1967. Flore et végétation des Monegros. *Botanica rhedonica*, sér. A, 3, p. 229-240.

EMBERGER, L. 1942. Un projet d'une classification des climats au point de vue phytogéographique. *Bull. Soc. Hist. nat. Toulouse*, 77, p. 97-124.

EMBERGER, L. 1951. Rapport sur les régions arides et semi-arides de l'Afrique du Nord. In: *Les bases écologiques de la régénération de la végétation des zones arides*, p. 50-61. International Union of Biological Sciences (IUBS), Paris.

EMBERGER, L. 1955. Une classification biogéographique des climats. *Recl. Trav. Labs. Bot. Zool. Univ. Montpellier*, 7, p. 3-43.

EMBERGER, L. 1959. La place de l'Australie méditerranéenne dans l'ensemble du vieux monde. In: *Biogeography and ecology in Australia*. A. Keast, R. L. Croker and C. S. Christian (Eds), p. 259-273. Monographiae Biologicae No. 8. W. Junk, The Hague.

ERINC, S.; TUNCDILEK, N. 1952. The agricultural regions of Turkey. *Geogrl. Rev.*, 42, p. 189-203.

EWELL, J. J.; MADRIZ, A. 1968. *Zonas de vida de Venezuela. Mapa ecologica (1/2 000 000), memoria explicativa*. Ministerio de Agricultura, Caracas.

FITZPATRICK, E. A. 1964. Seasonal distribution of rainfall in Australia analysed by Fourire methods. *Arch. Met. Geophy. Bioklim.*, ser. B, 13, p. 270-286.

FITZPATRICK, E. A.; SLATYER, R. O.; KRISHNAN, A. I. 1967. Incidence and duration of periods of plant growth in central Australia as estimated from climatic data. *Agricultural Meteorology*, 4, p. 389-404.

FOLEY, J. C. 1945. *Frost in the Australian region*. Bulletin No. 32. Commonwealth Meteorological Bureau, Melbourne.

FREITAG, H. 1971. Die natürliche Vegetation des südost-spanischen Trockengebietes. *Bot. Jahrb. Stuttgart*, 91 (2/3), p. 147-308.

GAO, SHAN-U, 1962. Sand fixation and afforestation in Jullin. *Scientia Sylvae*, 9 (2).

GARCIA, E. 1964. *Modificaciones al sistema de clasificación climatica de Koppen*. Inst. Geo. Univ. Nac. Mexico.

GARDNER, C. A. 1959. The vegetation of Western Australia. In: *Biogeography and ecology in Australia*. A. Keast, R. L. Crocker and C. S. Christian (Eds), p. 274-302. Monographiae Biologicae No. 8. W. Junk, The Hague.

GAUSSEN, H. et al. 1964. *Carte internationale du tapis végétal à 1/1 000 000 « Ceylon ». Notice explicative*. Institut français, Pondichéry.

GAUSSEN, H.; LEGRIS, P.; MEHER-HOMJI, V. M. 1972. *Carte internationale du tapis végétal à 1/1 000 000. «Rajasthan». Notice explicative*. Institut français, Pondichéry.

GENTILLI, J. 1953a. *Weather and climate in Western Australia*. Western Australian Government Tourist and Publicity Bureau.

GENTILLI, J. 1953b. Une critique de la méthode de Thorn-Twaite pour la classification des climats. *Annls. Geogr.*, 331, p. 180-185.

GENTILLI, J. 1971. Climates of Australia. In: *World survey of climatology*. Vol. 13. H. E. Lansberg (Ed.), p. 35-211. Elsevier, Amsterdam.

GENTILLI, J. 1972. *Australian climatic patterns*. Nelson, Melbourne.

GILLET, H. 1968a. Le peuplement végétal du massif de l'Ennedi (Tchad). *C. r. somm. Séanc. Soc. Biogéogr.*, 44 (383/388), p. 95-106.

GILLET, H. 1968b. Le peuplement végétal du massif de l'Ennedi (Tchad). *Mém. Mus. natn. Hist. nat., Paris*, sér. B, 17, 1-206.

GILLET, H. 1968c. Tchad et Sahel tchadien. In: *Conservation of vegetation in Africa south of the Sahara*. I. Hedberg and O. Hedberg (Eds), p. 54-58. Acta Phytogeographica Suecica 54. University of Uppsala, Uppsala.

GILLETT, J. B. 1948. *Provisional list of the trees and shrubs found in Iraq*. Iraq Department of Agriculture, Botanical Section Baghdad.

GOOD, R. d'O. 1954. The Bahrain islands and their desert flora. In: *Biology of deserts*. J. L. Cloudsley-Thompson (Ed), p. 45-55. Institute of Biology, London.

GUEST, E. 1932. Notes on trees and shrubs for tower Iraq. *Dept. Agric. Iraq Bull.*, 26, p. 1-18.

GUEST, E. 1933. Notes on plants and plant products with their colloquial names in Iraq. *Dept. Agric. Iraq Bull.*, 27, p. 1-111.

GUEST, E.; AL-RAWI, Ali. 1960. *Flora of Iraq, an account of the geology, soil, climate and ecology of Iraq, with gazetteer, glossary and bibliography.* vol. 1. Republic of Iraq, Ministry of Agriculture, Baghdad.

GUPTA, R. K.; PRAKASH, I. (Eds). 1975. *Environmental analysis of the Thar desert.* English Book Depot, Dehra Dun, India.

GWYNNE, M. D. 1968. Socotra. In: *Conservation of vegetation in Africa south of the Sahara.* I. Hedberg and O. Hedberg (Eds), p. 179-185. Acta Phytogeographica Suecica 54. University of Uppsala, Uppsala.

HARE, F. K. 1977. Climate and desertification. In: *Desertification: its causes and effects*, p. 63-167. Pergamon, Oxford. (Text compiled and edited by the secretariat of the United Nations Conference on Desertification.)

HILLS, E. S. (Ed.). 1966. *Arid lands, a geographical appraisal.* Methuen, London. Unesco, Paris.

HUBBLE, G. D. 1970. Soils. In: *Australian grasslands.* R. M. Moore (Ed.), p. 45-58. Australian National University Press, Canberra.

HUECK, K.; SEIBERT, P. 1972. *Vegetationskarte von Südamerika.* G. Fisher Verlag, Stuttgart.

HUMBERT, H.; COURS DARNE. 1965. *Carte internationale du tapis végétal «Madagascar» à 1/1 000 000. Notice explicative.* Institut français, Pondichéry.

INSTITUTO GEOGRAFICO "AGUSTIN CODAZZI". 1976. *Mapa ecológica de Colombia. 1/3 500 000.* Ministerio de Hacienda y crédito público, Bogota.

INSTITUTO MEXICANO DE RECURSOS NATURALES RENOVABLES. 1955. *Problemas de las zonas aridas de México.* Biblioteca Central de la Ciudad Universitaria, Mexico D.F.

IONESCO, T. 1965. Considérations bioclimatiques et phytogéographiques sur les zones arides du Maroc. *Cah. Rech. agron. Rabat*, 19, p. 1-130.

IVANOV, A.; VAKULIN, A. 1962. Osnovniye peskoukrepitelniye rasteniya pustin i polupustin venutrenni Mongolii. *Botanicheskyi Zhurnal*, 47, p. 1680-1684. (The main plants used for the fixation of shifting sands in the deserts and semi-deserts of Inner Mongolia.)

JAEGER E. C. 1956. *Desert wild flowers.* Stanford University Press, Stanford, California.

KACHKAROV, D. N.; KOROVINE, E. P. 1942. *La vie dans les déserts.* Payot, Paris.

KARSCHON, R. 1954. *La culture des Eucalyptus en Israël.* Sous-Commission de Coordination des Questions forestières méditerranéennes. IVe Session, Athènes. (Document FAO/CEF/SCM/46b).

KARSCHON, R. 1955. Techniques de reboisement en Israël. *J. For. Suisse*, 106, p. 215-221.

KARSCHON, R. 1961a. *Acacia albida* Del. in Israel and the Near East. *La-yaaran*, 11 (2), p. 4-7.

KARSCHON, R. 1964b. Contributions to the arboreal flora of Israel: *Acacia raddiana* Savi and *A. tortilis* Hayne. *La-yaaron*, 11 (3/4), p. 6-16.

KARSCHON, R. 1964a. The continentality of the climate of Israel. *La-yaaran*, 14 (4), p. 102-106 et 133-136.

KARSCHON, R. 1964b. Afforestation and tree planting in the arid areas of the Negev of Israel. *Ann. arid Zone*, 3, p. 13-24.

KASSAS, M. 1952. Habitat and plant communities in the Egyptian desert. I. Introduction. *J. Ecol.*, 40, p. 342-351.

KASSAS, M. 1953. Landforms and plant cover in the Egyptian desert. *Bull. Soc. Géogr. Egypte*, 24, p. 193-205.

KASSAS, M. 1962. Studies on the ecology of the Red Sea coastal land. *Bull. Soc. Géogr. Egypte*, 35, p. 129-175.

KASSAS, M.; GIRGIS, W. A. 1965. Habitat and plant communities in the Egyptian desert. VI. The units of a desert ecosystem. *J. Ecol.*, 53, p. 715-728.

KASSAS, M.; IMAN, M. 1954. Habitat and plant communities in the Egyptian desert. III. The wadi bed ecosystem. *J. Ecol.*, 424, p. 44-441.

KAUL, R. N. (Ed.). 1970. *Afforestation in arid zones.* Monograpiae biologicae No. 20. W. Junk, The Hague.

KEAY, R. W. J. 1959. *Vegetation map of Africa south of the Tropic of Cancer/Carte de la végétation de l'Afrique au sud du Tropique du Cancer.* Oxford University Press, London.

KEAST, A.; CROCKER, R. L.; CHRISTIAN, C. S. (Eds). 1959. *Biogeography and ecology in Australia.* Monographiae Biologae No. 8. W. Junk, The Hague.

KÖPPEN, W. 1936. Das geographische System der Klimate. In: *Handbuch der Klimatologie.* vol. 1, part C. W. Köppen and W. Geiger (Eds). Gebrüder Borntraeger, Berlin.

LABROUE, L.; LEGRIS, P.; VIART, M. 1965. Bioclimats du sous-continent indien. *Inst. fr. Pondichéry, Trav. Sect. scient. techn.*, 2, (3), p. 1-32. (3 cartes à 1/2 500 000).

LEEPER, G. W. (Ed.). 1970. *The Australian environment.* CSIRO - Melbourne University Press, Melbourne.

LEGRIS, P. 1963. La végétation de l'Inde : écologie et flore. *Inst. fr. Pondichéry, Trav. Sect. scient. techn.*, 6, p. 1-590.

LEGRIS, P.; VIART, M. 1961. Bioclimates of South India and Ceylon. *Inst. fr. Pondichéry, Trav. Sect. scient. techn.*, 3 (2), p. 165-178. (1 carte à 1/2 500 000).

LE HOUEROU, H. N. 1973. *Contribution à une bibliographie des phénomènes de désertification.* Rapport présenté au Colloque international sur la désertification, Nouakchott. FAO, Rome.

LEONT'EV, A. 1962. *Pescanve pustyni Svednej Azii i ih lesomeliorativnoe osvocnic.* Gosizdat, Uzb.SSR, Tashkent. (The sandy deserts of Soviet central Asia and their reclamation forests.)

LOGAN, R. F. 1960. *The central Namib Desert, South West Africa.* National Academy of Sciences - National Research Council, Publication 758. Washington.

LOGAN, R. F. 1969. Geography of the central Namib Desert. In: *Arid lands in perspective.* W. G. McGinnies, B. J. Goldman (Eds), p. 127-143, University of Arizona Press, Tucson.

MABBUTT, J. A. 1969. Landforms of arid Australia. In: *Arid lands of Australia.* R. O. Slatyer and R. A. Perry (Eds), p. 11-32, Australian National University Press, Canberra.

McGINNIES, W. G.; GOLDMAN, B. J.; PAYLORE, P. (Eds). 1968. *Deserts of the world: an appraisal of research into their physical and biological environments.* University of Arizona Press, Tucson.

MALLEUX, O. J. 1975. *Mapa forestal del Peru. Memoria explicativa.* Universidad nacional agraria, L Molina.

MANI, M. S. 1974. *Ecology and biogeography in India.* Monographiae Biologicae No. 23. W. Junk, The Hague.

MARIOLOPOULOS, E.; MISTARDIS, G.; CATACOUSINOS, D. 1964. Les régions semi-arides de la Grèce. In: *Land use in semi-arid Mediterranean climates/Utilisation des terres en climat semi-aride méditerranéen*, p, 59-61. Arid Zone Research/Recherches sur la zone aride 26. Unesco, Paris.

MARTONNE, E. de. 1935. Problèmes des régions arides sud-américaines. *Annls Géogr.,* 247, p. 1-27.

MEHER-HOMJI, V. M. 1963. Les bioclimats du sub-continent indien et leurs types analogues dans le monde. *Inst. fr. Pondichéry, Trav. Sect. scient. techn.,* 7 (1-2), p. 1-386.

MEHER-HOMJI, V. M. 1974. Variability and the concept of a probable climatic year. *Arch. Met. Geophys. Bioklim.,* ser. B, 22, p. 149-167.

MEIGS, P. 1952. World distribution of arid and semi-arid homoclimates. In: *Reviews of research on arid zone hydrology*, p. 203-210. Arid Zone Research 1. Unesco, Paris.

METRO, A. 1970. The Maghrib of Africa north of the Sahara. In: *Afforestation in arid zones.* R. N. Kaul (Ed.), p. 37-58. Monographiae Biologiae No. 20. W. Junk, The Hague.

MISTARDIS, G. 1962. Sur le paysage naturel de l'île Andros. *Pages andriotes,* 3, p. 17-20. (en grec).

MONOD, Th. 1951. Biologie et régions arides. In: *Les bases écologiques de la régénération de la végétation des zones arides*, p. 33-44. Union Internationale des Sciences Biologiques (UISB), Paris.

MONOD, Th. 1957. *Les grandes divisions chorologiques de l'Afrique.* Pub. No 24. Conseil scient. Afrique au Sud du Sahara.

MONOD, Th. 1971. Remarques sur les symétries floristiques des zones sèches Nord et Sud en Afrique. *Mitt. bot. St-Samml., Münch.,* 10, p. 375-423.

MONOD, Th. 1973. *Les déserts.* Horizons de France, Paris.

MONTERO DE BURGOS, J. L.; GONZALES REBOLLAR, J. L. 1974. *Diagramas bioclimaticas.* Instituto para la conservación de la naturaleza, Madrid.

MOORE, R. M. (Ed.). 1970. *Australian grasslands.* Australian National University Press, Canberra.

MORELLO, J. 1955-1956. Estudios botanicos en las regiones aridas de la Argentina. *Revta agron. NE Argent.,* 1, p. 301-370 et 385-524.

MOSINO, P. A. 1974. Los climas de la Republica Mexicana. In: *El escenario geografico*, p. 57-172. Inst. Nac. Antrop. Hist. Mexico, Mexico, D.F.

NIX, H. A.; AUSTIN, M. P. 1973. Mulga: a bioclimatic analysis. *Tropical Grasslands,* 7, p. 9-21.

NORLINDH, T. 1949. *Flora of the Mongolian steppe areas.* Reports from the scientific expedition to the north-western provinces of China. Tryckeri akiebolaget thule, Stockholm.

OBEID, M.; SEIF EL DIN, A. 1970. Ecological studies of the vegetation of the Sudan. I. *Acacia senegal* (L.) Willd. and its natural regeneration. *J. appl. Ecol.,* 7, p. 507-518.

OEDEKOVEN, K. H. 1970. The Near East (United Arab Republic, Iraq, Jordan, Kuwait, Yemen). In: *Afforestation in arid zones.* R. N. Kaul (Ed.), p. 86-135. Monographiae Biologicae No. 20. W. Junk, The Hague.

PENMAN, H. L. 1953. The physical bases of irrigation control. In: *Report 13th Int. Hort. Congr.,* 2, p. 913-924. Royal Horticultural Society, London.

PERRY, R. A. 1970. Arid shrublands and grasslands. In: *Australian grasslands.* R. M. Moore (Ed.), p. 247-259. Australian National University Press, Canberra.

PETROV, M. P. 1950. *Podvizhnye peski pustin Soyuza SSR i borda s mimi.* Geografgiz, Moscou. (Les sables mobiles des déserts de l'URSS et la lutte contre leur envahissement).

PETROV, M. P. 1952. *Agrolesomelioratsiya peskov v pustinyakh i polypustinayakh Soyuza SSR.* Académie des Sciences de la Turkménie, Achkhabab. (La phytoamélioration des sables dans les déserts et semi-déserts de l'URSS).

PETROV, M. P. 1957. La phytoamélioration des déserts de sable en URSS. *Annls Géogr.,* 66, (357), p. 397-410.

PETROV, M. P. 1970a. The USSR. In: *Afforestation in arid zones.* R. N. Kaul (Ed.), p. 210-233. Monographiae Biologicae No. 20. W. Junk, The Hague.

PETROV, M. P. 1970b. Central Asia. In: *Afforestation in arid zones.* R. N. Kaul (Ed.), p. 234-267. Monographiae Biologicae No. 20. W. Junk, The Hague.

PHILIPPIS, A. de. 1970. Italy. In: *Afforestation in arid zones.* R. N. Kaul (Ed.), p. 1-20. Monographiae Biologicae No. 20, W. Junk, The Hague.

PLANHOL, X. de; ROGNON, P. 1970. *Les zones tropicales arides et subtropicales.* Armand Colin, Paris.

POUQUET, J. 1963. *Les déserts.* Que sais-je ? No. 500. Presses universitaires de France, Paris.

PUIG, H. 1976. *Végétation de la Huasteca, Mexique.* Mission archéol. française au Mexique.

QUEZEL, P. 1964. 1968. *Carte internationale du tapis végétal et des conditions écologiques : feuilles « Largeau » et « Djado » à 1/1 000 000.* ICITV, Faculté des Sciences, Toulouse.

QUEZEL, P. 1971. Die Pflanzenwelt. 1. Teil: Flora und Vegetation der Sahara. In: *Die Sahara und ihre Randgebiete,* I. H. Schiffers (Ed.), p. 429-475, München.

RADWANSKI, S. A.; WICKENS, G. E. 1967. The ecology of *Acacia albida* on mantle soils in Zalingei Jebel Marra, Sudan. *J. appl. Ecol.,* 4, p. 569-579.

RAHEJA, P. C. 1966. Aridity and salinity. In: *Salinity and aridity. New approaches to old problems.* H. Boyko (Ed.), p. 43-130. Monographiae Biologicae No. 16. W. Junk, The Hague.

REPARAZ, G. 1958. La zone aride du Pérou. *Geogr. Annls.,* 40, p. 1-62.

RESEARCH COUNCIL OF ISRAEL. 1953. *Desert research.* Proceedings of an international symposium held in Jerusalem in May 1952. Special Publication No. 2. Research Council of Israel.

RIQUIER, J.; ROSSETTI, Ch. 1976. *Considérations méthodologiques sur l'établissement d'une carte des risques de désertification.* Rapport d'une consultation technique. FAO, Rome.

ROQUERO DE LABURU, C. 1964. L'utilisation du sol dans la région semi-aride de l'Espagne. In: *Land use in semi-arid Mediterranean climates/Utilisation des terres en climat semi-aride méditerranéen*, p. 75-80. Arid Zone Research/Recherches sur la zone aride 26. Unesco, Paris.

RZEDOWSKI, J. 1973. Geographical relationships of the flora of Mexican dry regions. In: *Vegetation and vegetational history of northern Latin America*. A. Graham (Ed.), p. 61-72. Elsevier, Amsterdam.

SAUVAGE, Ch. 1963. Le quotient pluviothermique d'Emberger, son utilisation et la représentation géographique de ses variations au Maroc. *Annls Serv. Phys. Globe Mét.*, Casabl., 20, p. 11-23.

SAUVAGE, Ch.; IONESCO, T. 1962. Les types de végétation du Maroc. Essai de nomenclature et de définition. *Revue Géogr. maroc.*, 1-2, p. 75-86.

SCHMITHÜSEN, J. 1956. Die rämliche Ordnung der chilenischen Vegetation. *Bonner Geogr. Abh.*, 17, 1-89.

SCHNELL, R. 1976. *Introduction à la phytogéographie des pays tropicaux*. Vol. 3-4. La flore et la végétation de l'Afrique tropicale. Gauthier-Villars, Paris.

SEARS, P. B. 1951. Biological problems involved in the agricultural use of arid regions. In: *Les bases écologiques de la régénération de la végétation des zones arides*, p. 45-49. International Union of Biological Sciences (IUBS), Paris.

SHREVE, F. 1942. The desert vegetation of North America. *Bot. Rev.*, 8, p. 195-246.

SHREVE, F.; WIGGINS, I. 1964. *Vegetation and flora of the Sonoran Desert*. Stanford University Press, Stanford, California.

SLATYER, R. O.; PERRY, R. A. (Eds). 1969. *Arid lands of Australia*. Australian National University Press, Canberra.

SPATE, O. H. K.; LEARMONTH, A. T. A. 1967. *India and Pakistan*. Methuen, London.

STACE, H. C. T.; HUBBLE, G. D.; BREWER, P.; NORTHCOTE, K.H.; SLEEMAN, J. R.; MULCAHY, M. J.; HALLSWORTH, E. G. 1968. *A handbook of Australian soils*. Rellim Technical Publications, Adelaide.

STANDLEY, P. C. 1920-26. *Trees and shrubs of Mexico*. U.S. National Herbarium, Contributions 23, Washington.

STEPHENS, C. G. 1961. *The soil landscapes of Australia*. Soil Publ. No. 18. CSIRO, Melbourne.

STEWART, G. A. 1959. Some aspects of soil fertility. In: *Biogeography and ecology in Australia*. A. Keast, R. L. Crocker and C. S. Christian (Eds), p. 303-314. Monographiae Biologicae No. 8. W. Junk, The Hague.

STOECKELER, J. H. 1970. The United States of America. In: *Afforestation in arid zones*. R. N. Kaul (Ed.), p. 268-346. Monographiae Biologicae No. 20. W. Junk, The Hague.

STOECKELER, J. H.; LIMSTROM, G. A. 1942. Ecological factors influencing reforestation in northern Wisconsin. *Ecol. Monogr.*, 12, p. 191-212.

SUDAN SURVEY DEPARTMENT. 1961. *Vegetation map of Sudan*. Khartoum. (1/4 000 000).

TACKHOLM, V. 1954. *Flora of Egypt*. Cairo University Press, Cairo.

TANOGLU, A. 1961. *Atlas of Turkey*. Faculté des Lettres, Université d'Istamboul, Istamboul.

THERON, A.; VINDT, J. 1955. *Carte de la végétation du Maroc « Rabat-Casablanca » à 1/200 000*. ICITV, Faculté des Sciences, Toulouse.

THORNTHWAITE, C. W. 1941. *Atlas of the climatic types in the United States*. Miscell. Publ. No. 421. U.S. Department of Agriculture, Forest Service.

THORNTHWAITE, C. W. 1948. An approach toward a rational classification of climate. *Geogrl Rev.*, 38, p. 55-94.

TOSI, J. A. 1960. *Zonas de vida natural en el Peru*. Bol. Techn. 5, Inst. interm. Ciencias agric. de la OEA.

TRICART, J. 1966. Un chott dans le désert chilien: la Pampa Tamarugal. *Revue Géomorph. dyn.*, 16, p. 12-22.

TRICART, J. 1969a. Le salar de Huasca. *Revue Géomarph. dyn.*, p. 49-84.

TRICART, J. 1969b. *Le modèle des régions sèches*. Traité de Géomorphologie 4. Société d'Edition d'Enseignement Supérieur (SEDES), Paris.

TROCHAIN, J. L. 1940. *Contribution à l'étude de la végétation du Sénégal*. Mémoires, No. 2. Institut français d'Afrique Noire, Brazzaville.

TROLL, C. 1959. Die tropischen Gebirge. Ihre dreidimensionale klimatische und pflanzengeographische Zonierung. *Bonn. geogr. Abhandl.*, 25, p. 1-93.

TROLL, C. 1960. The relationship between the climates, ecology and plant geography of the Southern cold temperate zone and the tropical high mountains. *Proc. R. Soc. Lond.*, ser. B, 152, p. 529-532.

TROLL, C. 1967. Die klimatische und vegetationsgeographische Gliederung des Himalaya-Systems. *Khumbu Himal.*, 1, p. 353-388.

TUMERTEKIN, E. 1956a. Some observations concerning dry farming in arid regions of Turkey. *Rev. Geogr. Inst. Univ. Instambul*, 3, p. 19-30.

TUMERTEKIN, E. 1956b. Dry months and dry seasons in Turkey. *Rev. Geogr. Inst. Istanbul*. 3, p. 74-79.

UNEP-FAO-UNESCO-WMO. 1977. *Desertification map of the world*. Map prepared for the United Nations Conference on Desertification (29 August-9 September 1977). Compiled by FAO and Unesco. Document A/CONF.74.2. UNEP, Nairobi.

UNESCO. 1958. *Climatology. Reviews of research*. Arid zone Research 10. Unesco, Paris.

UNESCO. 1964. *Land use in semi-arid Mediterranean climates/Utilisation des terres en climat semi-aride méditerranéen*. Arid Zone Research/Recherches sur la zone aride 26. Unesco, Paris.

UNESCO-FAO. 1963. *Bioclimatic map of the Mediterranean zone. Explanatory notes*. Arid zone research 21. Unesco, Paris.

UNESCO-FAO. 1970. *Carte de la végétation de la région méditerranéenne. Notice explicative/Vegetation of the Mediterranean zone. Explanatory notes*. Recherches sur la zone aride/Arid Zone Research 30. Unesco, Paris.

U.S. DEPARTMENT OF AGRICULTURE. 1941. *Climate and man*. U.S. Department of Agriculture Yearbook. Washington, D.C.

U.S. DEPARTMENT OF AGRICULTURE. 1949. *Trees*. U.S. Department of Agriculture Yearbook. Washington, D.C.

U.S. DEPARTMENT OF AGRICULTURE, 1955. *Water*. U.S. Department of Agriculture Yearbook. Washington, D.C.

U.S. DEPARTMENT OF AGRICULTURE. 1957. *Soil*. U.S. Department of Agriculture Yearbook. Washington, D.C.

U.S. DEPARTMENT OF AGRICULTURE. 1958. *Land*. U.S. Department of Agriculture Yearbook. Washington, D.C.

VIDAL de la BLACHE, P.; GALLOIS, L. 1929. *Géographie universelle*. Armand Colin, Paris.

VOLK, O. H. 1964. Die afro-meridional-occidentale Floren-Region in Südwestafrika. In: *Beiträge zur Phytologie*. Kreeb (Ed.), p. 1-16. Stuttgart.

VOLK, O. H. 1966. Einfluss von Mensch und Tier auf die natürliche Vegetation im tropischen Südwest-Afrika In: *Beiträge zur Landespflege,* 2 (1/2), p. 108-131.

WADHAM, S.; WILSON, R. K.; WOOD, J. 1964. *Land utilisation in Australia*. Melbourne University Press, Melbourne.

WALTER, H. 1973. *Die Vegetation der Erde*. Stuttgart.

WALTER, H.; LIETH, H. 1960. *Klimadiagram Weltatlas*. G. Fischer Verlag, Jena.

WILLIAMS, C. B. 1954. Some climatic observations in the Egyptian desert. In: *Biology of deserts*. J. L. Cloudsley-Thompson (Ed.), p. 18-27. Institute of Biology, London.

YARON, B.; DANFORD, E.; VAADIA, Y. (Eds). 1973. *Arid zone irrigation*. Ecological Studies No. 5. Springer-Verlag, Berlin.

YASSOGLOU, N. J.; CATACOUSINOS, D.; KOUSKOLEKAS, A. 1964. Land use in the semi-arid zone of Greece. In: *Land use in semi-arid Mediterranean climates/Utilisation des terres en climat semi-aride méditerranéen*, p. 63-67. Arid Zone Research/Recherches sur la zone aride 26. Unesco, Paris.

ZAMORA, C. J. 1971. *Mapa de capacidad de uso de los suelos del Perú*. Oficina Nacional de Evaluación de Recursos Naturales (ONERN), Lima.

ZOHARY, M. 1944. Vegetational transects through the desert of Sinai. *Palest. J. Bot.*, 3, p. 57-78.

ZOHARY, M. 1962. *Plant life of Palestine*. Ronald Press, New York.

ZOHARY, M. 1973. *Geobotanical foundations of the Middle East*. 2 vol. Gustav Fischer Verlag, Stuttgart.

Unesco publications: national distributors (Abridged list)

Argentina	EDILYR, S.R.L., Tucumán 1699 (P.B. 'A'), 1050, BUENOS AIRES.
Australia	*Publications:* Educational Supplies Pty. Ltd., Post Office Box 33, BROCKVALE 2100, N.S.W. *Periodicals:* Dominie Pty. Subscriptions Dept., P.O. Box 33, BROOKVALE 2100, N.S.W. *Sub-agent:* United Nations Association of Australia (Victorian Division), 2nd floor, Campbell House, 100 Flinders St., MELBOURNE 3000.
Austria	Dr. Franz Hain, Verlags-und Kommissionsbuchhandlung, Industriehof Stadlau, Dr. Otto-Neurath-Gasse 5, 1220 WIEN.
Brazil	Fundação Getúlio Vargas, Editora-Divisão de Vendas, caixa postal 9.052-ZC-02, Praia de Botafogo 188, RIO DE JANEIRO, R.J.
Burma	Trade Corporation no. (9), 550-552 Merchant Street, RANGOON.
Canada	Renouf Publishing Company Ltd., 2182 St. Catherine Street West, MONTREAL, Que., H3H 1M7.
Chile	Bibliocentro Ltda., Constitución n° 7, Casilla 13731, SANTIAGO (21).
Cuba	Instituto Cubano del Libro, Centro de Importación, Obispo 461, LA HABANA.
Czechoslovakia	SNTL, Spalena 51, PRAHA 1. (Permanent display): Zahranicni literatura, 11 Soukenicka, PRAHA 1. *For Slovakia only:* Alfa Verlag Publishers, Hurbanova nam. 6, 893 31 BRATISLAVA.
Denmark	Ejnar Munksgaard Ltd., 6 Norregade, 1165 KOBENHAVN K.
Ecuador	RAYD de publicaciones, Garcia 420 y 6 Diciembre, casilla 3853, QUITO; Casa de la Cultura Ecuatoriana, Núcleo del Guayas, Pedro Moncayo y 9 de Octubre, casilla de correos 3542, GUAYAQUIL.
Ethiopia	Ethiopian National Agency for Unesco, P.O. Box 2996, ADDIS ABABA.
Finland	Akateeminen Kirjakauppa, Keskuskatu 1, SF-00100 HELSINKI 10.
France	Librairie de l'Unesco, place de Fontenoy, 75700 PARIS. CCP 12598-48.
German Democratic Republic	Buchhaus Leipzig, Postfach 140, 701 LEIPZIG, or international bookshops in the German Democratic Republic.
Germany Federal Republic	S. Karger GmbH, Karger Buchhandlung, Angerhofstrasse 9, Postfach 2, D-8034 GERMERING/MUNCHEN. *For scientific maps only:* Geo Center, Postfach 800830, 7000 STUTTGART 80. *For 'The Courier' (German edition only):* Colmantstrasse 22, 5300 BONN.
Ghana	Presbyterian Bookshop Depot Ltd., P.O. Box 195, ACCRA; Ghana Book Suppliers Ltd., P.O. Box 7869, ACCRA; The University Bookshop of Ghana, ACCRA; The University Bookshop of Cape Coast; The University Bookshop of Legon, P.O. Box 1, LEGON.
Greece	International bookshops (Eleftheroudakis, Kauffmann, etc.).
Hong Kong	Swindon Book Co., 13-15 Lock Road, KOWLOON; Federal Publications (HK) Ltd., 5a Evergreen Industrial Mansion, 12 Yip Fat Street, Wong Chuk Hang Road, ABERDEEN.
Hungary	Akadémiai Könyvesbolt, Váci u. 22, BUDAPEST VI.
Iceland	Snaebjörn Jonsson & Co., H.F., Hafnarstraeti 9, REYKJAVIK.
India	Orient Longman Ltd., Kamani Marg, Ballard Estate, BOMBAY 400 038; 17 Chittaranjan Avenue, CALCUTTA 13; 36a Anna Salai, Mount Road, MADRAS 2; B-3/7 Asaf Ali Road, NEW DELHI 1; 80/1 Mahatma Gandhi Road, BANGALORE-560001; 3-5-820 Hyderguda, HYDERABAD-500001. *Sub-depots:* Oxford Book & Stationery Co., 17 Park-Street, CALCUTTA 700016; Scindia House, NEW DELHI 110001; Publications Section, Ministry of Education and Social Welfare, 511, C-Wing, Shastri Bhavan, NEW DELHI 110001.
Indonesia	Bhratara Publishers and Booksellers, 29, Jl. Oto Iskandardinata 111, JAKARTA; Gramedia Bookshop, Jl. Gadjah Mada 109, Indira P.T., Jl. Dr. Sam Ratulangi 37, JAKARTA PUSAT.
Iran	Iranian National Commission for Unesco, Avenue Iranchahr Chomali No. 300, B.P. 1533, TEHRAN; Krarazmie Publishing and Distribution Co., 28 Vessal Shirazi Street Avenue, P.O. Box 314/1486, TEHRAN.
Ireland	The Educational Company of Ireland Ltd., Ballymount Road, Walkinstown, DUBLIN 12.
Israel	Emanuel Brown, formerly Blumstein's Bookstores, 35 Allenby Road, and 48 Nachlat Benjamin Street, TEL AVIV; 9 shlomzion Hamalka Street, JERUSALEM.
Jamaica	Sangster's Book Stores Ltd., P.O. Box 366, 101 Water Lane, KINGSTON.
Japan	Eastern Book Service Inc., C.P.O. Box 1728, TOKYO 10092.
Kenya	East African Publishing House, P.O. Box 30571, NAIROBI.
Republic of Korea	Korean National Commission for Unesco, P.O. Box Central 64, SEOUL.
Libyan Arab Jamahiriya	Agency for Development of Publication and Distribution, P.O. Box 34-35, TRIPOLI.
Madagascar	Commission Nationale de la République Démocratique de Madagascar pour l'Unesco, Boîte postale 331, TANANARIVE.
Malaysia	Federal Publications Sdn. Bhd., Lot 8238 Jalan 222, Petaling Jaya, SELANGOR.
Malta	Sapienzas, 26 Republic Street, VALLETTA.
Mexico	SABSA, Insurgentes Sur n° 1032-401, MEXICO 12, D.F.
Netherlands	N.V. Martinus Nijhoff, Lange Voorhout, 9, 's-GRAVENHAGE; Systemen Keesing, Ruysdaelstraat 71-75, AMSTERDAM 1007.
Netherlands Antilles	Van Dorp-Eddine N.V., P.O. Box 200. Willenstad, CURAÇAO, N.A.
Nigeria	The University Bookshop of Ife; The University Bookshop of Ibadan, P.O. Box 286; The University Bookshop of Nsukka; The university Bookshop of Lagos; The Ahmadu Bello University Bookshop of Zaria.
Norway	Publications: Johan Grundt Tanum, Kari Johans gate 41/43, OSLO 1. *For 'The Courier':* A/S Narvesens Litteraturtjeneste, Box 6125, OSLO 6.
Pakistan	Mirza Book Agency, 65 Shahrah Quaid-e-asam, P.O. Box 729, LAHORE 3.
Philippines	The Modern Book Co., 926 Rizal Avenue, P.O. Box 632, MANILA D-404.
Southern Rhodesia	Textbook Sales (PVT) Ltd., 67 Union Avenue, SALISBURY.
Singapore	Federal Publications (S) Pte. Ltd., No. 1 New Industrial Road, off Upper Paya Lebar Road, SINGAPORE 19.
South Africa	Van Schaik's Bookstore (Pty) Ltd., Libri Building, Church Street, P.O. Box 724, PRETORIA.
Spain	MUNDI-PRENSA LIBROS S.A., Castelló 37, MADRID 1; Ediciones LIBER, Apartado 17, Magdalena 8, Ondárroa (Vizcaya); DONAIRE, Ronda de Outeiro, 20, apartado de correos 341, La Coruña; Libreria AL-ANDALUS, Roldana, 1 y 3, SEVILLA 4; LITEXSA, Libreria Técnica Extranjera, Tuset, 8-10 (Edificio Monitor), BARCELONA.
Sudan	Al Bashir Bookshop, P.O. Box 1118, KHARTOUM.
Sweden	*Publications:* A/B C.E. Fritzes Kungl. Hovbokhandel, Fredsgatan 2, Box 16356, 103 27 STOCKHOLM 16. *For 'The Courier':* Svenska FN-Förbundet, Skolgränd 2, Box 150 50, S-104 65 STOCKHOLM.
Switzerland	Europa Verlag, Rämistrasse 5, 8024 ZURICH; Librairie Payot, 6, rue Grenus, 1211 GENEVA 11.
Thailand	Suksapan Panit, Mansion 9, Rajdammern Avenue, BANGKOK; Nibondh & Co. Ltd., 40-42 Charoen Krung Road, Siyaeg Phaya Sri, P.O. Box 402, BANGKOK; Suksit Siam Company, 1715 Rama IV Road, BANGKOK.
Uganda	Uganda Bookshop, P.O. Box 145, KAMPALA.
U.S.S.R.	Mezhdunarodnaja Kniga, MOSKVA G-200.
United Kingdom	H.M. Stationery Office, P.O. Box 569, LONDON SE1 9NH; Government Bookshops: London, Belfast, Birmingham, Bristol, Cardiff, Edinburgh, Manchester.
United Republic of Tanzania	Dar es Salaam Bookshop, P.O. Box 9030, DAR ES SALAAM.
United States	Unipub, 345 Park Avenue South, New York, N.Y. 10010.

A complete list of distributors is available from the Office of Publications, Unesco [4]